English Language Edition Consultant:
I. Aleksander, Brunel University, England

An Introduction to ROBOT TECHNOLOGY

Philippe Coiffet and Michel Chirouze

Kogan Page

Translated by Meg Tombs

First published 1982 by Hermes Publishing (France)
51 Rue Rennequin, 75017 Paris, France

Copyright © Hermes Publishing (France) 1982
English language edition first published 1983
by Kogan Page Ltd, 120 Pentonville Road, London N1 9JN
English language translation copyright © Kogan Page Ltd 1983
All rights reserved

British Library Cataloguing in Publication Data
Coiffet, Philippe and Chirouze, Michel
   Introduction to Robotics
     1. Cybernetics
     I. Title II. Elements de Robotique. *English* III.
   ISBN 0 85038 637 3
Printed and bound in Great Britain by Redwood Burn Ltd

Contents

Foreword   9

Chapter 1   **Robotics: an introduction**   11
Areas of application of robotics, 12
The new industrial revolution, 14
The aim of this book, 15

Chapter 2   **Robots and robots in use**   17
Origin of the word 'robot', 17
Definition of the word 'robot', 17
Characteristic properties of a robot, 17
General structure of a robot, 18
The robot environment, 19
Task description, 20
Role of the computer, 20
Typical industrial robot, 21
Classification of robots, 21
Robot generations, 22
Existing robots and the robot market, 23

Chapter 3   **Representation of a robot**   25
Functional representation, 25
Graphical representation, 27
Arms: structures in use, 29
Structure of end effectors, 29

Chapter 4   **Degrees of freedom of a robot**   33
Degrees of freedom of a rigid object, 33
Degrees of freedom of a robot, 34
Degrees of freedom specific to a tool, 35
Degrees of freedom and mobility, 35

Chapter 5   **Basic principles of control**   37
Variables to be handled, 37
The main levels of control, 39

Chapter 6   **Control based on the geometrical model**   43
Geometrical model: a much simplified robot model, 43
Geometrical or positional control, 51

Chapter 7	**Control based on the kinematic model**	57

Kinematic model: a simplified robot model, 57
Variational control, 60
Characteristics of kinematic control, 64
Models and dynamic control, 65

Chapter 8	**Actuator servocontrol**	67

Principles of servocontrol, 67
Mathematical study of a servo-system, 75
Specific practical problems involved in the use of a robot servo-system, 79

Chapter 9	**Robot actuators**	83

Pneumatic actuators, 84
Hydraulic actuators, 86
Servocontrolled hydraulic systems, 89
Electrical actuators, 91
Servocontrolled electrical motors, 96
Transmission systems, 103
Conclusions, 104

Chapter 10	**Internal sensors**	107

Movement or position sensors, 107
Speed sensors, 115
Stress sensors, 116
Acceleration sensors, 119

Chapter 11	**External sensors**	121

Applications of external sensors, 121
Tactile sensors, 122
Stress sensors, 124
Proximity sensors, 125
Visual sensors, 126

Chapter 12	**Computer control**	129

Analog-digital, digital-analog converters, 130
Other types of converter used in robotics, 130
The program, 131
Conclusions, 134

Chapter 13	**Robot training and trajectory generation**	137

Methods of recording trajectories, 137
Manual control used in training, 140
Trajectory generation, 142
Trajectories in the task space and in the articulated variable space, 144
Control languages, 145
Conclusions, 146
References, 146

Chapter 14	**Robot performance and standards**	147

What is robot performance?, 147
Task performance, 147
Human performance in robot control, 150
Economic performance, 150
Performance standards, 151

Chapter 15	**Robots in use**	153
	Examples of uses, 153 End effector components, 159 Conclusions, 163	
Appendix I	**Matrix calculations in robotics: a summary**	167
	Use of matrix calculations, 167 Handling real term matrices: a summary, 168	
Appendix II	**Mathematical summary: transformation of cooordinate sets**	173
	Components of a vector in an orthogonal set of coordinates, 173 Transformation of coordinate set, 174 Specific examples useful for modelling and control of robots, 175 Inverse transformation, 177	
Appendix III	**Summary of the principles of hydraulic flow**	179
	Definitions and equations, 179	
Appendix IV	**Direct current motors**	181
	Working principles, 181 Motor with induction control, 182 Motor with armature control, 183	
Appendix V	**The dynamic model**	185
	Problems associated with dynamic control, 185 Dynamic control, 186 Effects of gravitational force, 187	
Bibliography		191
Index		193

Foreword

Robotics is now a well established field of endeavour both in industry and research laboratories. There is a danger that the word may be widely used even in areas where it is inappropriate, so knowing precisely what a robot is, how it is controlled and how it may be used in specific applications is of the highest importance.

The authors are not only innovators in the development of robots but also highly respected educators. This book has been carefully compiled to crystallize, for the reader, the fundamentals of robot operation and application. The material carefully treads its path between achieving broad coverage and depth where it is needed. Industrialists, teachers and students alike will benefit from the book.

Igor Aleksander
July 1983

Chapter 1
Robotics: an introduction

As a result of the great advances of the last few years many industrial processes have become largely automated, with the human operator playing an ever decreasing role. The fully automated and unmanned factory is probably now only a few decades away.

Although the idea of total automation is not new, it was not believed to be practicable until less than ten years ago. The study of automatic control and the use of automation systems dates back to the Second World War. But it was some years later that there was a huge leap forward in the design of automation systems. This comparatively sudden development marks the time at which the first industrial *robot* appeared. It was then that the term *robotics* started to be used to describe a new academic and industrial discipline.

The evolution of automation systems into robotic systems has taken place in two stages, at least as far as industrial applications are concerned (see Figure 1.1).

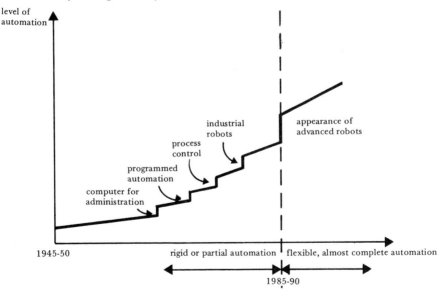

Figure 1.1. *Graph showing the development of robotics*

In the first stage, the distinguishing features of robotic systems, those that set them apart from existing automated machines, were their versatility and flexibility. It was thought that a large range of physical tasks, particularly repetitive ones, could be entrusted to robots, and that a large number of existing machines, as well as human operators, would be replaced by the robot. Such robots, and most of those in use at present fall into this category, were developed as a form of programmed automation. However, they could not react to their environments, and, more importantly, to *changes* in their environments. We are now witnessing the beginning of the second stage of development, that of the *interactive robot*, which can apprehend and react to changes in its physical environment.

The need for such new, sophisticated devices is based on the following considerations:
1. There are limits to the range of application of programmable robots, and this would be much enhanced by the introduction of sensing and other devices.
2. Many tasks performed by human operators are said to be repetitive, but are not so in every detail. Slight changes in the environment require constant adjustments in the processes involved. The robot must then be able to take account of such changes in the environment in order to replace the human operator.
3. The robot, although a highly sophisticated device, cannot by itself solve all the problems that can be solved by the human operator, and must therefore be associated with additional techniques, for example computer-aided manufacture (CAM).

Robotics now has two currently accepted meanings:
1. in the strictest sense of the word, it implies the further development of automation by improving the robot as we know it at present;
2. in the broader sense of the word, it involves the development of not only the robot itself, but also processes associated with the robot, for example, computer-aided design (CAD), or the consideration of the robot as a machine with special properties in association with other machines (eg CAM, flexible workshops).

1.1 Areas of application of robotics

There are three main areas in which robots can be used, but these areas should not be thought of as necessarily separate as there is considerable overlap between them. The areas are *production, exploration* and *aids for the handicapped*.

1.1.1 PRODUCTION

Manufacturers have used robots mostly to reduce manning levels. Robots, used either with other robots or with other machines, have two major advantages compared with traditional machines. They allow:

1. almost total automation of production processes, leading to a higher quality of end product, better quality control and an increased adaptability to varying demand;
2. adaptability, at speed, of the production unit, allowing the production line to be switched from one product to another similar product, for example, from one model of a car to another, or when a breakdown immobilizes one element in the production unit.

Adaptable production units such as those mentioned in (2) above are known as *flexible manufacturing systems*. A *flexible unit* comprises a small number of robots and machines used together (eg a robot designed to use a lathe, together with the lathe itself, would be called a flexible unit). A number of flexible units used together is called a *flexible workshop*. The same expression is used when a number of machines and robots are used together.

1.1.2 EXPLORATION

During *exploration* operations, work is carried out in hostile environments, for example, under water, in outer space, in radioactive environments or in high-temperature environments. Therefore, robots can be employed either as *autonomous robots* or in *teleoperation systems*.

1.1.2.1 The autonomous robot

Robots of this kind when placed in a hostile environment perform pre-programmed tasks. For example, robots could be used to collect rocks from the surface of Mars, inspect nuclear reactors, mine the sea bed and assist in foundries. At present, autonomous robots are employed only for simple tasks and cannot be expected to 'think', ie they are unable to interpret their environment for themselves. It is for this reason that, in practice, *teleoperation* systems are mostly used in exploration work.

1.1.2.2 Teleoperation

This method involves placing a robot (known as a *slave*) in a hostile environment, whilst a human operator directs the machine by remote control. Handling radioactive materials by remote control is a well known use of this technique in practice. Because the operator controls the slave's movements, information about changes in the slave's environment must be transmitted to the operator. The most straightforward method employed is to transmit a visual image using cameras, but it is also necessary to monitor the mechanical strain exerted on the slave. This is known as a *sensory feedback system*.

1.1.3 AIDS FOR THE HANDICAPPED

Robots of more or less human form are often depicted in films, helping or replacing man in unpleasant, difficult or dangerous tasks. Robots of this sort belong to science fiction at present. The nearest that could be envisaged today is a household robot that can clean the floor whilst avoiding the furniture, but the cost would be prohibitive and it is of limited interest. However, in medical robotics, developments in aids for the individual have led to great improvements in the living conditions of paralytics (paraplegics and tetraplegics) and amputees. Medical robotics has been used for:

Prostheses: artificial hands and legs etc.
Ortheses: rigid motorized structures placed around paralysed limbs to train their movements.
Teletheses: for use by tetraplegics. Then, robots controlled by the handicapped person, using the parts of the body in which voluntary mobility has been maintained (tongue, mouth, eye muscles etc), are employed.

1.2 The new industrial revolution

If, in twenty or thirty years time, the production of consumer goods and the running of services were largely automated, the economic, social and political consequences would be far-reaching. As stated at the beginning of this chapter, there was a significant leap forward in the development of automation systems which was marked by a realization that almost complete 'robotization' of certain industrial tasks was a real possibility. This realization had both a technical basis and an economic rationale.

1.2.1 TECHNICAL BASIS

1. Accelerated development in the technology necessary for 'robotization' particularly connected with data-processing, microelectronics and automation;
2. the appearance of new devices possessing new properties (eg

adaptable robots associated with CAD and CAM systems).

1.2.2 ECONOMIC BASIS

Investment in the research and development necessary to robotics was favoured by the economic situation in the industrialized countries. The economic crisis has drawn attention to the cost of manpower, and automation was seen as a means of improving productivity. Future investment is likely to favour the development of robotics rather than traditional industrial work methods.

1.3 The aim of this book

In this book robotics is considered in its narrowest sense, describing and discussing only the device known as the robot, and not the enlarged systems in which the robot can be used. It is written for:

1. the automation scientist, who develops models and different forms of control, which are in fact mathematical algorithms;
2. the manufacturer, who assembles the constituent parts of a robot;
3. the user, who will teach the robot the tasks it must perform and use it to increase productivity and improve working conditions for human workers;
4. the teacher who wishes to introduce students in engineering and science to the foundations of robotics.

Chapter 2
Robots and robots in use

2.1 Origin of the word 'robot'

The fact that the word 'robot' exists in many languages is evidence of its recent coinage, despite the fact that it represents the fulfilment of an age-long aspiration: to create a device that can replace man in everything he cannot or does not wish to do himself, but still needs or wants to do, without presenting any threat to his authority.

The term first came into use during the 1920s and 1930s, following the appearance of a play by the Czech author Karel Capek, called *R.U.R.* (*Rossum's Universal Robots*). In the play small, artificial and anthropomorphic creatures strictly obeyed their master's orders. These creatures were called 'robots', a word derived from the Czech *robota*, meaning 'forced labour'.

2.2 Definition of the word 'robot'

Two possible definitions are:

1. The definition supplied by the *Concise Oxford Dictionary*:
   'Apparently human automaton, intelligent and obedient but impersonal machine'.
   This definition cannot, however, be entirely accurate since no existing robot in use resembles a human being, nor is intended to.
2. The definition supplied by the Robot Institute of America:
   'A reprogrammable and multifunctional manipulator, devised for the transport of materials, parts, tools or specialized systems, with varied and programmed movements, with the aim of carrying out varied tasks'.

This definition, although better, is still far from perfect.

2.3 Characteristic properties of a robot

In addition to seeking a definition, it is useful to describe the charac-

teristic features of a robot. Two principal properties emerge:
1. *Versatility*: The structural/mechanical potential for performing varied tasks and/or performing the same task in different ways. This implies a mechanical structure with variable geometry. All existing robots have this quality.
2. *Auto-adaptability to the environment*: This complex expression simply means that a robot must be designed to achieve its objective (the performance of a task) by itself, despite unforeseen, but limited, changes in the environment during the performance of the task. This ability of some industrial robots described in this book requires that the robot be aware of its environment, ie have artificial senses. Robot senses so far developed are modest compared with the human capacity for interpreting the environment, but intensive research is being carried out in this area.

2.4 General structure of a robot

The operational robot is made up of five interactive elements (see Figure 2.1). These are:
1. *The articulated mechanical system (AMS)* which comprises the actual limbs — arms, joints and end effectors — brought together in an interdependent and interconnected array. In Chapters 6 and 7 the behaviour and relevant mathematical modelling, which can be complex, describing the behaviour of such systems are outlined.
2. *The actuators* which provide power, under carefully controlled conditions, for the AMS. The power can be electrical, hydraulic or pneumatic and each of these methods for energizing robot articulations is considered in Chapter 9.
3. *The transmission devices or systems* which link the actuators with the AMS, producing movements in the individual articulations with power provided by the actuators. Such systems may take the form of cables, bands, notched belts, gears etc (see Chapter 9).
4. *The sensors* which are the 'eyes' of the robot. These may take the form of tactile, electrical or optical devices, amongst others, and are used to obtain information on the position of the articulations and on objects in the robot's environment. Such sensors are said to be *proprioceptive,* or *internally sensed,* when they determine the positions of individual articulations and the configuration of the whole system, relative to a reference position known as the *initialization configuration.* The configuration of the AMS then determines the form of the robot.
5. *The robot 'brain' or computer* is the unit that processes data received by the robot through its sensors. In less sophisticated

robots, it is merely a machine with a fixed program while in more advanced robots, the 'brain' takes the form of a fully programmable digital minicomputer or microprocessor.

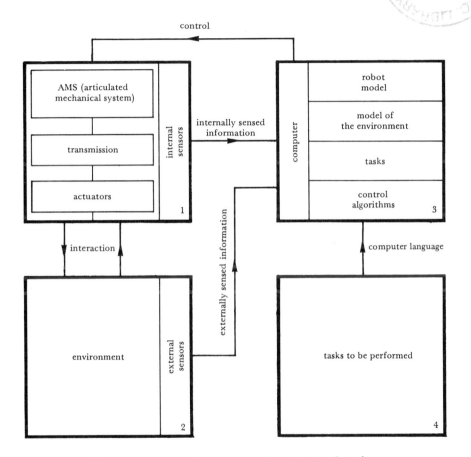

Figure 2.1. *Schematic representation of robot function*

2.5 The robot environment

These are the surroundings in which the robot is placed and operates. For a robot in a fixed position, the environment is the attainable space defined by the volume covered by all the possible configurations of the end effector.

In the environment, the robot encounters *obstacles* that are to be avoided, and *objects of interest* upon which it must act. Thus *interaction* is established between the robot and the environment. For example, if a robot is to manipulate objects placed on a table, it must take into account the presence of the table, since, if it attempts to pass through the table, its movements will be opposed. Thus the *action* of the robot (passage through the table) is opposed by the *reaction* of the table (offering resistance).

2.5.1 ENVIRONMENTAL SENSORS

Information on the state of the environment is gathered by *external* or *exteroceptive* sensors (particularly information concerning the location of external objects). Cameras, force detectors, proximity sensors and tactile sensors can be used for the acquisition of this information.

2.6 Task description

It is necessary to describe the desired task in terms that the computer can understand. Depending on the system used, the form of the language can be gestural, oral or written.

2.7 Role of the computer

In robots, the role of the brain is taken by the computer. In its memory are stored:
1. *a model of robot behaviour*, ie the relationship between the excitation signals of the actuator and the consequential movements of the robot;
2. *a model of the environment*, ie a description of everything within its attainable space, for example, the zones in which it must not function because of obstacles;
3. *programs (i)* that allow the computer to understand the tasks to be performed;
4. *programs (ii)* that provide control of the robot structure, so that it can carry out what is required. The sequences of computer instructions are known as *control algorithms*.

During the performance of a task, the computer:
1. assesses the state of the robot, using *internally sensed* information;
2. assesses the state of the environment, using *externally sensed* information;
3. uses its collection of stored models and recorded programs and generates commands (ie power signals to the actuators) which cause the structural robot to proceed towards the successful completion of the task required of it.

2.8 Typical industrial robot

A typical industrial robot (see Figure 2.2) (at least 98 per cent of all robots in use are of this type) comprises the structural robot (ie the system of articulations) and the computer. The robot is not designed to understand its environment (ie it has no external sensors). This means that the computer could be replaced by a simpler programmed automaton which carries out specific tasks irrespective of changes in the environment. Because the computer does not possess a model of the environment, but simply controls a series of pre-programmed movements, it is necessary to foresee any changes that may take place in the environment to compensate for this.

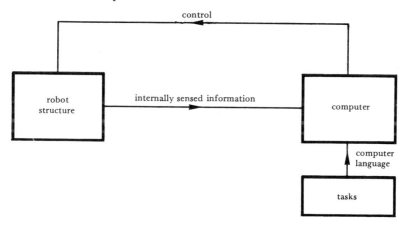

Figure 2.2. *Schematic representation of a typical industrial robot*

In practice, such robots can be used if the tasks to be performed are strictly repetitive. Because the means of sensing the environment — automatic environmental analysis (*scene analysis*) — is only partly developed, it is costly and not wholly reliable. However, industrial robots equipped with limited-performance visual systems or strain-measurement systems are now being produced.

2.9 Classification of robots

Real robots possess, in varying degrees, the two characteristic properties mentioned before: versatility and auto-adaptability to their environment. In order to classify robots, it is necessary to be able to define and distinguish between various types, but this has given rise to a certain confusion. There are three different modes of classification in current use, that adopted by the *Japanese Industrial Robot Association (JIRA)*, that in use at the *Robotics Institute of America (RIA)* and the one used by the *Association Française de Robotique Industrielle (AFRI)*.

2.9.1 THE JIRA CLASSIFICATION

Robots are divided into six classes:

Class 1: manual handling device: a device with several degrees of freedom (DOF) actuated by the operator;

Class 2: fixed sequence robot: handling device which performs the successive stages of a task according to a predetermined, unchanging method, which is difficult to modify;

Class 3: variable sequence robot: the same type of handling device as in Class 2, but the stages can be modified easily;

Class 4: playback robot: the human operator performs the task manually by leading or controlling the robot, which records the trajectories. This information is recalled when necessary, and the robot can perform the task in the automatic mode;

Class 5: numerical control robot: the human operator supplies the robot with a movement program rather than teaching it the task manually;

Class 6: intelligent robot: a robot with the means to understand its environment, and the ability to successfully complete a task despite changes in the surrounding conditions under which it is to be performed.

2.9.2 THE RIA CLASSIFICATION

In this classification only the machines in Classes 3, 4, 5 and 6 of the JIRA system are considered to be robots.

2.9.3 THE AFRI CLASSIFICATION

Type A: Class 1: handling device with manual control or telecontrol;

Type B: Classes 2 and 3: automatic handling device with a predetermined cycle;

Type C: Classes 4 and 5: programmable, servoed robot (continuous or point-to-point trajectory), known as *first generation robots*;

Type D: Class 6: acquisition of certain data on the environment, known as *second generation robots*.

2.10 Robot generations

In the JIRA classification all robots with the ability to apprehend their environment are grouped in Class 6, which will thus incorporate future robot designs. The AFRI classification is more cautious, and only includes in Class 6 existing robots with the ability to acquire specific data on their environment. For the time being the range of this data remains limited. This is the reason for the division of robots into different generations: the first generation is made up of Classes 4 and 5, the second generation is made up of part of Class 6 and the third and following generations will be made up of the robots of the future, which will possess properties and characteristics as yet hardly understood or at present poorly controlled (eg three-dimensional vision, comprehension of natural language).

2.11 Existing robots and the robot market

The robot market consists of robots from Classes 3 to 6. The first industrial robot was sold in about 1961 by Unimation. According to the RIA, in 1981 there were approximately 20,000 robots in operation world-wide.

The distribution was:

Japan	14,000	France	1,000
USA	4,100	Sweden	600
USSR	3,600	Great Britain	500
West Germany	2,300		

In the USA, the use of industrial robots can be broken down as follows:

foundry work:	between 615 and 840
spot welding:	between 1,435 and 1,500
machine loading:	between 820 and 850
materials handling:	between 540 and 615
assembly:	between 400 and 410
other:	between 205 and 400

Estimated numbers of robots in use in 1985 are about 65,000 in Japan; 50,000 in the USA and 30,000 in Western Europe.

Predicted statistics for the USA are as follows:

— market value: 1980, $M100; 1990, $M2,000;
— robots in use with a CAD/CAM system: 1985, 10 to 12 per cent; 1990, 20 to 25 per cent;
— robots equipped with a vision system: 1981, 1 per cent; 1990, 25 per cent.

The growth in the value of the market may reach 90 per cent per year, which is considerable. In 1982, there were 100 producers of Class 3 to 6 robots in Japan and over 50 in the USA.

Chapter 3
Representation of a robot

A system can be represented by making a model of it. The nature of the model can vary greatly depending upon what is to be shown: it can take the form of an oral description, mathematical equations or drawings etc. In this chapter the functional and graphical representation of the AMS (articulated mechanical system) will be discussed.

3.1 Functional representation

The geometrical structure of the robot system can be broken down into three functional, interconnected sub-groups (see Figure 3.1):

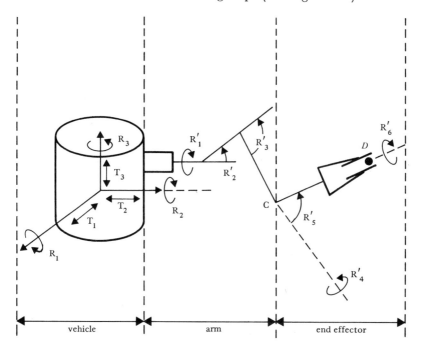

Figure 3.1. *Functional representation of a robot with 12 DOF: nine rotational $(R_1, R_2, R_3, R'_1, R'_2, R'_3, R'_4, R'_5$ and $R'_6)$ and three translational $(T_1, T_2$ and $T_3)$. D is the gripper*

1. *The vehicle*: If a robot is to perform tasks in a particular location, it must first reach this location, and this is achieved by the *vehicle* — in the form of a terrestrial machine, underwater device or satellite. The vehicle will possess between two and six degrees of freedom (DOF) (see Chapter 4), depending on the environment in which it operates. For example, Figure 3.2 shows a mounted manipulator. It has six DOF: three *translational* (T), used to change its position, and three *rotational* (R), used to change its orientation.

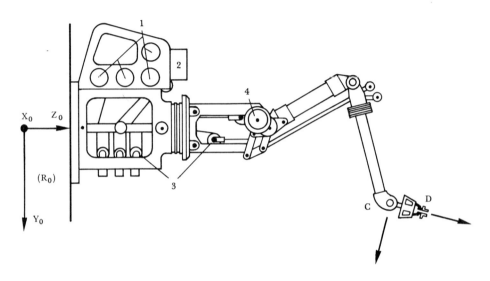

Figure 3.2. *Side view of the manipulator robot MA23 (CEA/La Calhène); 1, motors; 2, servocontrol system; 3, transmission systems; 4, counterbalancing device; D, gripper*

2. *The arm*: The arm manoeuvres the end effector (the gripper D, or rather point C in Figures 3.1 and 3.2) to a precise location. This cannot generally be performed by the vehicle alone. Moreover, fixed robots, which constitute the majority of industrial robots, have no vehicle, by definition. The positioning of C is carried out using the three DOF, rotations R'_1, R'_2 and R'_3 shown in Figure 3.1.

3. *The end effector*: This is the tool, for example a gripper, used in manipulators. If point C is accurately specified, the movement of the end effector depends only on its orientation. Thus, three possible independent rotations are required (R'_4, R'_5 and R'_6).

Sometimes, however, when very precise movements are required, the end effector can rotate around the three normal axes so that minute adjustments to the position of D can be made.

This schematic division into vehicle, arm and end effector often corresponds to the real use of mechanical structures. Nevertheless, the articulations of the arm and end effector are sometimes coupled, which complicates control and invalidates the concept of division into subgroups.

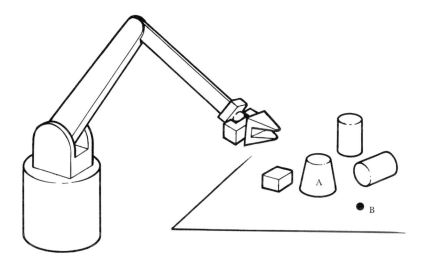

Figure 3.3. *Diagram of a robot in perspective*

3.2 Graphical representation

To draw a robot in side view or in perspective, as in Figures 3.2 and 3.3 respectively, is complicated and does not give a clear picture of how the

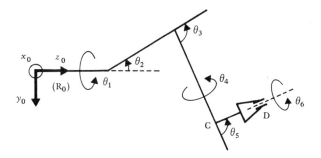

Figure 3.4. *Geometrical structure of the manipulator robot shown in Figure 3.2. The first three DOF (θ_1, θ_2 and θ_3) allow point C, and thus the gripper (D), to be positioned in space. The last three DOF (θ_4, θ_5 and θ_6) allow the gripper to be oriented in any direction*

Type of articulation	Relative movements	Number of DOF	Symbols
Fixed beam	0 rotation 0 translation	0	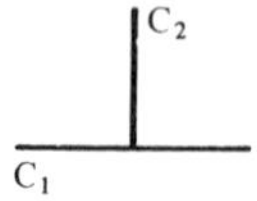 C_1: body 1 C_2: body 2
Pin	1 rotation 0 translation	1	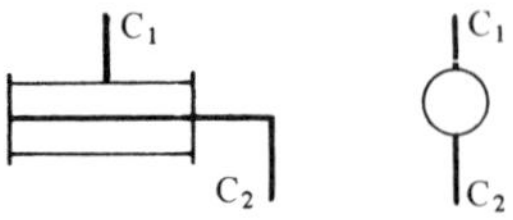
Slide	0 rotation 1 translation	1	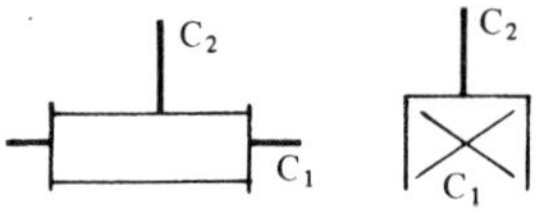
Helical slide	1 rotation 1 coupled translation	1	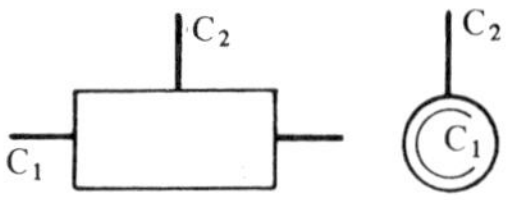
Sliding pin	1 rotation 1 translation	2	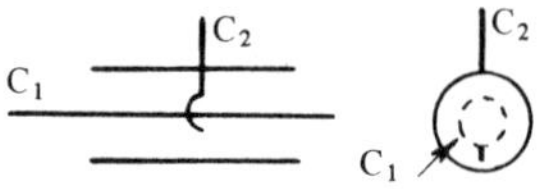
Sill	1 rotation 2 translations	3	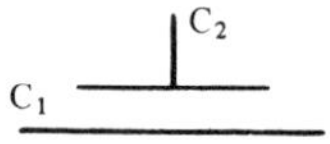
Swivel	3 rotations 0 translation	3	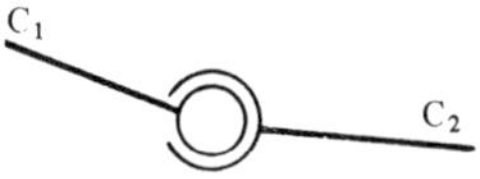
Linear joint	2 rotations 2 translations	4	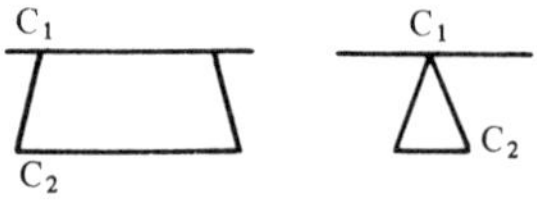
Round joint	3 rotations 1 translation	4	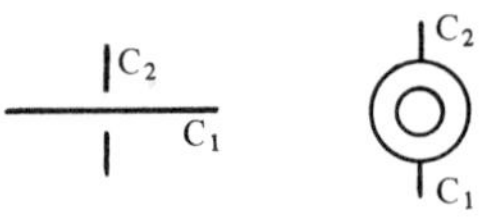
Contact point	3 rotations 2 translations	5	
Free joint	3 rotations 3 translations	6	No symbol No contact between bodies

Figure 3.5. *Representation of mechanical articulations*

various segments move in relation to each other. For this reason it is better to draw a skeletal diagram as shown in Figure 3.4. To ensure easy understanding a representational standard should be referred to, particularly when describing the types of articulation, ie the mechanical relationships between segments used.

A typical form of representation for different types of articulation is shown in Figure 3.5.

3.3 Arms: structures in use

Many manufacturers offer robots which consist of an arm for which the user can choose a particular tool. For these arms, which incoporate the first three DOF of the fixed robot, the number of potential combinations of rotations and translations constituting the articulations is limited. In practice, only four or five different arm structures currently are used. These are shown in Figure 3.6.

A translation representing a parallel mechanical relationship is termed P. A rotation is termed R. Thus, an arm can be identified as follows (see Figure 3.6):

— RRR or 3R for class 1;
— PRR or P2R for class 2;
— RRP or 2RP for class 3;
— PRP or RPP or R2P for class 4;
— PPP or 3P for class 5.

3.4 Structure of end effectors

The function of the arm is to position the end effector or tool correctly. The main function of the end effector is to orient the tool. There are generally three possible rotations of the end effector about three perpendicular axes. It is very important that the three axes should intersect at one point, for two reasons:

— it allows the tool to be oriented without any need for further translation;
— the robot becomes resolvable (see Chapter 7).

This means that the robot can be controlled easily when in position.
There are exceptions to this design:

1. Many industrial robots have only one or two rotational DOF in their end effector.

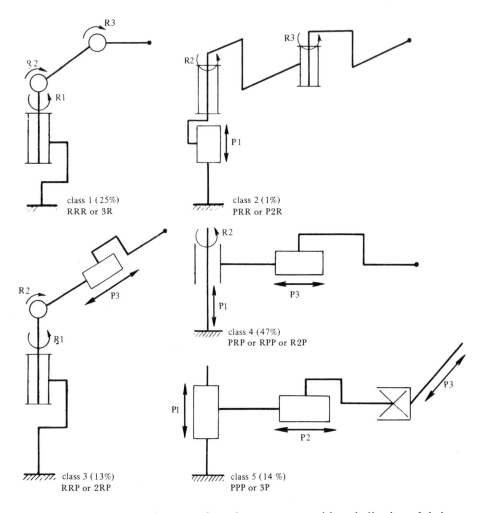

Figure 3.6. *The five main types of arm in current use, with an indication of their percentage use in parentheses. A translation representing a parallel mechanical relationship is termed P and a rotation is termed R*

2. There are often delicate technical problems in the swivel-limb relationship.
3. The intended application sometimes requires another structure, such as the addition of a translational DOF.

Examples:

— the three axes of rotation are normal and intersect two by two (see Figure 3.7);
— the three axes of rotation are normal and do not intersect (see Figure 3.8);

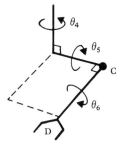
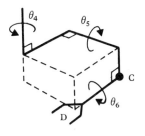

Figure 3.7. *Structure of end effector, with axes intersecting two by two*

Figure 3.8. *Structure of end effector without intersection of the three axes of rotation*

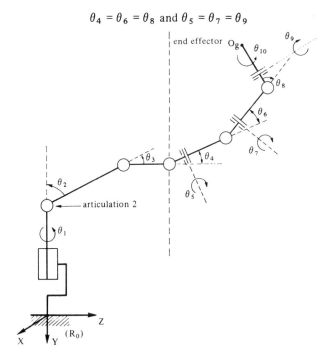

Figure 3.9. *The skeleton of a robot having ten articulations but possessing only six DOF:* $\theta_4 = \theta_6 = \theta_8$ *and* $\theta_5 = \theta_7 = \theta_9$ *(structure of the ACMA painter robot)*

— the axes are not normal and are not fixed. This is the case in Figure 3.9. The end effector of the 'painter robot' has only three DOF: θ_4, θ_5 and θ_{10}, since $\theta_4 = \theta_6 = \theta_8$ and $\theta_5 = \theta_7 = \theta_9$ will be maintained. This structure is used, for example, in paint spraying. Such an arrangement is needed to bring the spraygun into position inside the body of the car.

Chapter 4
Degrees of freedom of a robot

4.1 Degrees of freedom of a rigid object

Any point on an object is related to an orthogonal set of coordinate axes. The number of independent movements the object can make with respect to the coordinate set R is called its number of *degrees of freedom* (DOF). The movements that the object can perform are:

- the three translations T_1, T_2 and T_3 along the axes OX, OY and OZ;
- the three rotations R_1, R_2 and R_3 about the axes OX, OY and OZ (see Figure 4.1).

This means that, by using these three translations and three rotations, the object can be oriented and moved with respect to the coordinate set R.

A simple object has six DOF. When a relationship is established between two objects, each loses DOF with respect to the other. A relationship of this kind can also be expressed in terms of the rotations and/or translations between the two bodies which have become impossible as a result of the relationship established.

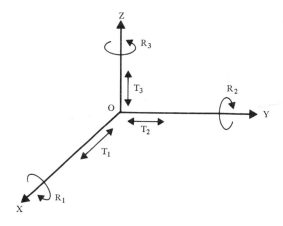

Figure 4.1. *The six DOF of a rigid object: three translational (T_1, T_2 and T_3) and three rotational (R_1, R_2 and R_3)*

4.2 Degrees of freedom of a robot

A robot is expected to move its end effector, or the tool attached to it, to a given point, with correct orientation. If the use to which the robot will be put is not known in advance it should be equipped with six DOF, but if the tool itself has a special structure, six DOF may not be required. For example, to position a sphere at a given point in space, three DOF are sufficient (see Figure 4.2). Five DOF are sufficient to position and orient a rotating drill which can be represented as a cylinder rotating about its principal axis (see Figure 4.3).

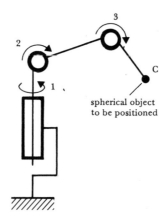

Figure 4.2. *Three DOF are needed to position a sphere at a given point in space*

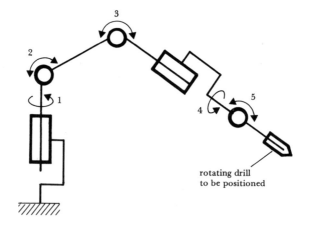

Figure 4.3. *Five DOF are required to position a rotating drill*

Generally, the arm of the robot is equipped with three DOF and any increase may be made in the end effector (see Figure 4.4).

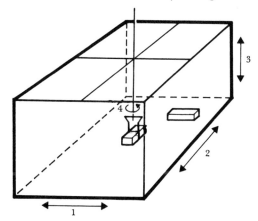

Figure 4.4. *Task requiring only four DOF: three translational (1, 2 and 3) and one rotational (4)*

4.3 Degrees of freedom specific to a tool

A robot might be required to drill holes. To do this its drill must rotate. The rotation is always powered by an ancillary motor, and therefore is not counted as a DOF of the robot.

The same applies for a manipulator robot. The gripper must be able to open and close. The DOF, which is specific to the gripper, cannot be counted as one of the DOF of the robot since it contributes to the operation of the gripper only. This is an important point.

4.4 Degrees of freedom and mobility

Degrees of freedom cannot be described as belonging to something with respect to something else. Thus, in Figure 4.5, point A has no DOF with respect to the fixed base, point B has two DOF and point C has three DOF with respect to the base. If the position of point D is specified, articulation C, which is used to move D, is theoretically superfluous, although this would not be so in practice necessarily. Articulation C can be considered to possess no longer a DOF, but a *degree of mobility*.

If, however, CD is to be oriented by positioning point C, articulation C becomes a DOF, allowing CD to be oriented within certain limits (two other DOF would be necessary to allow CD to take up any orientation).

It must be remembered that:

1. Not all mobility constitutes a DOF. An articulation may become a

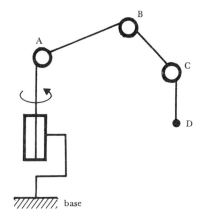

Figure 4.5. *Point A has no DOF with respect to the base, point B has two DOF and point C three DOF with respect to the base*

DOF in the context of the function it performs, but not invariably (see Figure 4.6).

2. A robot never uses more than six independent DOF, but may use many more degrees of mobility.

This distinction is very important in establishing control of a robot (degrees of freedom may be redundant).

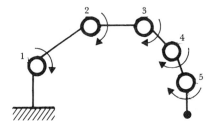

Figure 4.6. *Despite the large number of articulations, this robot does not have more than two independent DOF under any circumstances*

Chapter 5
Basic principles of control

Control of the robot is developed in its 'brain' or *processor*. Depending on the case, this can take any of a variety of forms (eg simple programmable automaton, minicomputer, microprocessor). Its structure can also be highly varied, from a single processor to a hierarchical, decentralized system using several processors, each one fulfilling a specific function or being in direct contact with a specific part of the robot (eg a degree of freedom or DOF, also frequently called an axis) (see Figure 5.1).

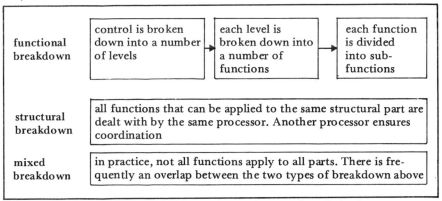

Figure 5.1. *Main methods of control breakdown and analysis. A processor can be used for any combination of the above functions*

The following section, without going into structural detail, is a discussion of the theory of the role of the processor.

5.1 Variables to be handled

Consider a robot of the type shown in Figure 5.2. If it is instructed to grasp piece A, it is vital to know the state of the end effector at any moment (ie its position, orientation, whether it is open or closed) relative to A. The position of A is known in a set of coordinate axes linked to the table on which it rests. This set of axes is called the *task axes* (R_0). The state of the end effector is expressed in this set of axes

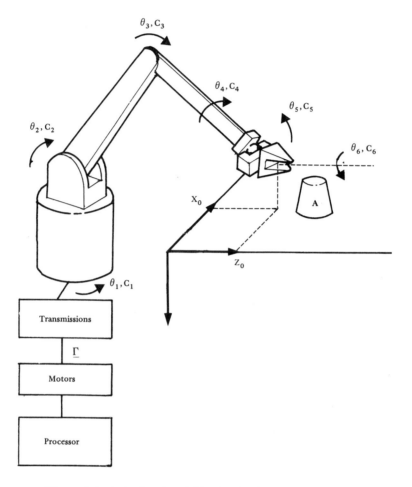

Figure 5.2. *Articulated variables involved in robot control*

by a number of values or parameters, which are the components of a vector $\underline{X}$. The problem is to control this vector in time, ie $\underline{X}(t)$. $\underline{X}$ is only changed because of movements of the articulations θ_1 to θ_6, which are considered together in the vector $\underline{\Theta}(t)$.

The articulations move because of the couples C_1 to C_6 which are developed in them. These couples together form the vector $\underline{C}(t)$ and are derived from torques $\underline{\Gamma}(t)$ delivered by the motors and sent to the articulations through the transmissions. The motors deliver the torque because they are powered using currents or voltages assembled into a vector $\underline{V}(t)$, controlled by the processor or processors.

Controlling a robot is essentially controlling the bidirectional equation:

$$\underline{V}(t) \leftrightarrow \underline{\Gamma}(t) \leftrightarrow \underline{C}(t) \leftrightarrow \underline{\Theta}(t) \leftrightarrow \underline{X}(t) \qquad (5\text{-}1)$$

5.2 The main levels of control

The main levels of control are shown in Figure 5.3.

5.2.1 LEVEL 1: ARTIFICIAL INTELLIGENCE

If a robot is given the command 'Go and fetch piece A!', how can the task be carried out? It should be established first of all that if the command is successfully executed it is because the robot is capable, at the very least, of generating the vector $\underline{X}(t)$ which represents the movement of the end effector relative to A (ie in the set of coordinate axes R_0 in which A has been located) for the command.

What takes place between presenting the robot with a command and the generation of $\underline{X}(t)$ constitutes the first, and highest, level of control. It involves all the problems associated with artificial intelligence: understanding of words, of natural language, plan generation and task description.

This level is still largely at the research stage, and will not be discussed further because it goes far beyond the scope of this introduction to robotics, and it is hardly if ever used in current industrial practice. The normal procedure is to give the elements concerning $\underline{X}(t)$, $\underline{\Theta}(t)$, directly to the robot in the training phase.

5.2.2 LEVEL 2 OR THE 'CONTROL MODE' LEVEL

This is the level at which bidirectional relationships can be established between $\underline{X}(t)$ and torques $\underline{\Gamma}(t)$ supplied by the motors or actuators. It is important to remember that there are several modes of control available. This is because the relationships:

$$\underline{X}(t) \leftrightarrow \underline{\Theta}(t) \leftrightarrow \underline{C}(t) \leftrightarrow \underline{\Gamma}(t) \qquad (5\text{-}2)$$

pose various problems in practice, and thus in order to achieve a satisfactory approach various hypotheses have been formed which can be very different to each other. They depend on the depth of knowledge the human operator possesses of the subject and the use to which the robot is to be put.

Consider equation (5-2). The relationships between the four vectors constitute models:

$\underline{\Gamma}(t)$	$\underline{C}(t)$	$\underline{\Theta}(t)$	$\underline{X}(t)$ (5-3)
model of transmissions	model of the robot as an articulated mechanical system (AMS)	model of the relationships between the articulated variables and the values to be controlled in the task space	model of the robot in real space

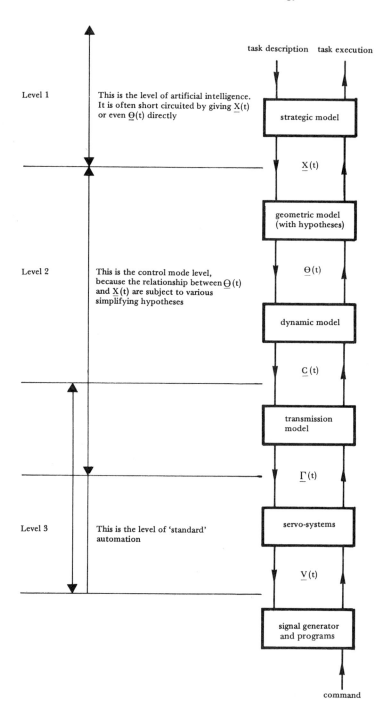

Figure 5.3. *The main levels of robot control*

The first problem involves the dynamics of the system. There are various difficulties that arise in attempting this:

1. There is no way of knowing how to model the imperfections in the articulations correctly (ie dry friction, flexibility in the segments of the articulations).
2. Even if it seems possible to take these into account, the model will include hundreds of parameters and the processor will be unable to perform all the necessary in-line operations at an adequate speed.
3. Control corresponds to an inversion of the model [if the desired $\underline{X}(t)$ is known, generate $\underline{\Gamma}(t)$ to correspond with it]. There is no doubt that the more complex the model, the more difficult this inversion, and all the more so when the model is non-linear.

This is why, in industry, sophisticated, complex models are not used, and two kinds of control (with many variants) are generally applied, corresponding to two types of model. These are based on static theory (the robot, in the course of its movements, passes through a succession of balanced states) and are the *geometrical model*, which makes use of the transformation of coordinates between $\underline{X}$ and $\underline{\Theta}$, and the *kinematic model*, which is simply a linearization of the geometrical model, and assumes that $\underline{X}$ and $\underline{\Theta}$ are simply subjected to small increases. Chapters 6 and 7 deal with these two types of control, and the problems of dynamic modelling and control are discussed further in Appendix V.

5.2.3 LEVEL 3 OR THE SERVO-SYSTEM LEVEL

This concerns the standard practices of robotics. The main points will be discussed later (see Chapters 8 and 9). Two points, however, should be noted:

1. The division between Levels 1 and 2 is not always clear, in particular it is a matter of choice whether or not to include transmissions and step-down gears in Level 1, which should solve the problem of the relationships:

$$\underline{V} \leftrightarrow \underline{\Gamma} \qquad (5\text{-}4)$$

or:

$$\underline{V} \leftrightarrow \underline{\Gamma} \leftrightarrow \underline{C} \qquad (5\text{-}5)$$

The current tendency is research into motors with integral step-down gears, which can be mounted directly onto the robot articulations. This does, however, pose problems concerning inertial torque and reduction ratio, which are for the present inadequate.

2. Standard servo-systems are the same, but have been replaced more and more by digital control servo-systems.

Chapter 6
Control based on the geometrical model

6.1 Geometrical model: a much simplified robot model

6.1.1 INTRODUCTION

A robot is made up of a succession of solid segments, mobile with respect to one another. This is called an *articulated mechanical chain*.

To control the movements of this articulated mechanical chain, which is equipped with motors, it is important to have available a representation of it in the form of mathematical equations, that is a model. The computer that controls the movements uses these mathematical models to predict what will happen.

It is virtually impossible to form an exact model since the mechanical parts are very complex. They may bend under pressure from a load, the articulations may be elastic, and they are subject to friction which is difficult to calculate. Usually approximate models are made, which although fairly simple, are of considerable use.

When planning these models two assumptions are made, that the segments of the robot are infinitely rigid, and that all the articulations are perfect, without friction or play.

To make the model simpler still it will be supposed also that the link between two successive segments only involves one DOF, either a perfect rotation or a perfect translation. When this is not possible, in, for example, a swivel joint with three independent DOF, imaginary bodies with zero mass and length are added. An example is shown in Figure 6.1.

6.1.2 REPRESENTATION OF AN ARTICULATED MECHANICAL CHAIN

6.1.2.1 Associated coordinate sets

Consider a robot with $n + 1$ successive segments, either real or imaginary (see Figure 6.2). A set of orthogonal coordinate axes is associated for each segment. The one related to the fixed segment is called the *fixed coordinate set (R_0)* or *coordinate task set*. The task is relative to this coordinate set.

To simplify the calculations, the vertex of each coordinate set is placed at the junction of the segments where the origin of the

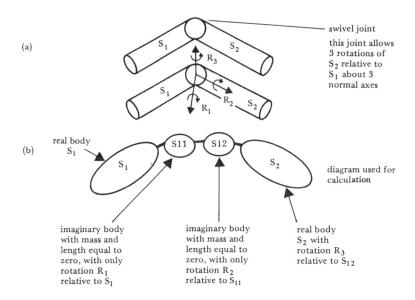

Figure 6.1. *Simplified model with links possessing a single DOF (a). When this is not possible, imaginary bodies of zero dimension and mass are added, having a single DOF (b)*

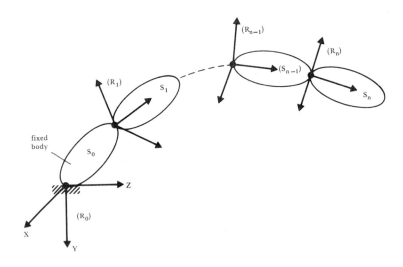

Figure 6.2. *Articulated chain and associated coordinate sets (R)*

translation, or the centre of rotation relative to the preceding segment is assumed to be. The coordinate sets are labelled so that each one (R_q) moves relative to the preceding one (R_q-1) with a rotation about an axis of $(R_q - 1)$ or with a translation along one of the axes of $(R_q - 1)$.

6.1.2.2 Conventional representation

When only the kinematic behaviour of the robot is to be considered (ie the positions and speeds of movement of the various articulations and the end effector) a conventional representation can be used.

Figure 6.3 is an example of the representation of a robot with four DOF, in which the nature of the links is clear, to facilitate choice of the associated coordinate link.

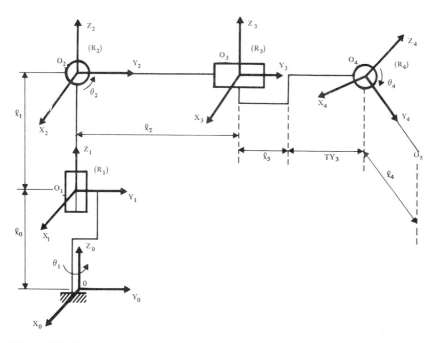

Figure 6.3. *Representation of robot with four DOF, for formulation of a model*

6.1.3 CALCULATION OF GEOMETRICAL MODEL

A fixed robot is made up of an *arm* which includes the first three DOF, and an *end effector* which interacts with the environment. The first point of interest is the orientation and position of the end effector, for example a gripper, in real space, in the fixed set of axes (R_0), known as the task axes. This can be calculated by taking Figure 6.3 as an example and using the information provided in Appendix II (see Section AII.3). The end effector is represented by the segment $O_4 O_5$ along Y_4. There is only one DOF, a rotation of coordinate set (R_4) about the axis X_3 of the preceding coordinate set (R_3) (which can be transferred to O_4 so as to be visualized better). It is necessary to calculate:

1. the coordinate set (R_4) (ie its unity vectors e_x^4, e_y^4, e_z^4) in the coordinate set (R_0) (ie as a function of e_x^0, e_y^0, e_z^0);
2. the coordinates of O_4 and O_5 in the coordinate set (R_0).

6.1.3.1 Orientation of (R_4)

In Appendix II it is proved that:

$$\frac{\text{base of }(R')}{\text{expressed in }(R)} = M_{R'}^R \times \frac{\text{base of }(R')}{\text{expressed in }(R')} \equiv M_{R'}^R \quad (6\text{-}1)$$

Thus in the present example, using simplified notation:

$$\text{base}(R_4/R_0) = M_{R_4}^{R_0} \equiv M_4^0 \quad (6\text{-}2)$$

From this formula, an interesting recurrence is derived:

$$\text{base}(R_n/R_{n-1}) = M_n^{n-1} \quad (6\text{-}3)$$

$$\text{base}(R_{n-1}/R_{n-2}) = M_{n-1}^{n-2} \quad (6\text{-}4)$$

but:

$$\text{base}(R_n/R_{n-2}) = M_n^{n-2} \quad (6\text{-}5)$$

where:

$$M_n^{n-2} = M_{n-1}^{n-2} M_n^{n-1} \quad (6\text{-}6)$$

(The order of the matrices is very important.) So:

$$\text{base}(R_4/R_0) = M_4^0 = M_1^0 M_2^1 M_3^2 M_4^3 \quad (6\text{-}7)$$

M_1^0 corresponds to a rotation of (R_1) relative to (R_0) through an angle of θ about Z_0.

$$M_1^0 = \begin{pmatrix} \cos\theta_1 & \sin\theta_1 & 0 \\ -\sin\theta_1 & \cos\theta_1 & 0 \\ 0 & 0 & 1 \end{pmatrix} \quad (6\text{-}8)$$

M_2^1 corresponds to a rotation of (R_2) relative to (R_1) through an angle of θ about X_1.

$$M_2^1 = \begin{pmatrix} 1 & 0 & 0 \\ 0 & \cos\theta_2 & \sin\theta_2 \\ 0 & -\sin\theta_2 & \cos\theta_2 \end{pmatrix} \quad (6\text{-}9)$$

M_3^2 corresponds to a translation of (R_3) relative to (R_2) of length TY_3 along the axis Y_2.

$$M_3^2 = \begin{pmatrix} 1 & 0 & 0 \\ 0 & 1 & 0 \\ 0 & 0 & 1 \end{pmatrix} = (\pi) \tag{6-10}$$

(compare with the result of translation: $\theta = 0$).

M_4^3 corresponds to a rotation of (R_4) relative to (R_3) through an angle of θ about X_3:

$$M_4^3 = \begin{pmatrix} 1 & 0 & 0 \\ 0 & \cos\theta_4 & \sin\theta_4 \\ 0 & -\sin\theta_4 & \cos\theta_4 \end{pmatrix} \tag{6-11}$$

To simplify the notations, let $\cos\theta_i \equiv Ci$ and $\sin\theta_i \equiv Si$. Thus:

$$M_4^0 = \text{base}\,(R_4/R_0) = \begin{pmatrix} C1 & S1 & 0 \\ -S1 & C1 & 0 \\ 0 & 0 & 1 \end{pmatrix}\begin{pmatrix} 1 & 0 & 0 \\ 0 & C2 & S2 \\ 0 & -S2 & C2 \end{pmatrix}\begin{pmatrix} 1 & 0 & 0 \\ 0 & C4 & S4 \\ 0 & -S4 & C4 \end{pmatrix}$$

$$M_4^0 = \begin{pmatrix} C1 & S1C(2+4) & S1S(2+4) \\ -S1 & C1C(2+4) & C1S(2+4) \\ 0 & -S(2+4) & C(2+4) \end{pmatrix} \tag{6-12}$$

This means that the precision of e_x^4 on the axes of (R_0) are $C1$ on OX_0, $S1C(2\mp4)$ on OY_0, $S1S(2\mp4)$ on $\overline{OZ_0}$ and that those of e_y^4 are $\underline{-S1}$ on OX_0 etc.

6.1.3.2 Coordinates of O_4 and O_5 in (R_0)

The following vector relationship may be written:

$$\underline{OO_5}(R_0) = \underline{OO_4}(R_0) + \underline{O_4O_5}(R_0) \tag{6-13}$$

but in Appendix II it is shown that:

$$\underline{V}(R) = M_R^{R'}\,\underline{O_4O_5}\,\underline{V}(R') \tag{6-14}$$

which is shown by:

$$\underline{O_4O_5}(R_0) = M_4^0\,\underline{O_4O_5}(R_4) \tag{6-15}$$

If $\underline{O_4O_5}$ is placed upon axis Y_4 the following can be written:

$$\underline{O_4O_5}(R_4) = \begin{pmatrix} 0 \\ \ell_4 \\ 0 \end{pmatrix} \begin{matrix} \leftarrow \text{component following } X_4 \\ \leftarrow \text{following } Y_4 \\ \leftarrow \text{following } Z_4 \end{matrix} \tag{6-16}$$

Thus, equation (6-13) is identical to:

$$\underline{OO_5}(R_0) = \underline{OO_4}(R_0) + M_4^0 \begin{pmatrix} 0 \\ \ell_4 \\ 0 \end{pmatrix} \quad (6\text{-}17)$$

Equation (6-17) is further broken down to give:

$$\underline{OO_4}(R_0) = \underline{OO_3}(R_0) + \underline{O_3O_4}(R_0) \quad (6\text{-}18)$$

$$\underline{O_3O_4}(R_0) = M_3^0 \underline{O_3O_4}(R_3) \quad (6\text{-}19)$$

$$\underline{O_3O_4}(R_3) = \begin{pmatrix} 0 \\ \ell_3 + TY_3 \\ 0 \end{pmatrix} \quad (6\text{-}20)$$

and this can be continued in the same way:

$$\underline{OO_3}(R_0) = \underline{OO_2}(R_0) + \underline{O_2O_3}(R_0) \quad (6\text{-}21)$$

$$\underline{O_2O_3}(R_0) = M_2^0 \underline{O_2O_3}(R_2) \quad (6\text{-}22)$$

$$\underline{O_2O_3}(R_2) = \begin{pmatrix} 0 \\ \ell_2 \\ 0 \end{pmatrix} \quad (6\text{-}23)$$

resulting in:

$$\underline{OO_2}(R_0) = \underline{OO_1}(R_0) + \underline{O_1O_2}(R_0) \quad (6\text{-}24)$$

$$\underline{O_1O_2}(R_0) = M_1^0 \underline{O_1O_2}(R_1) \quad (6\text{-}25)$$

$$\underline{O_1O_2}(R_1) = \begin{pmatrix} 0 \\ 0 \\ \ell_1 \end{pmatrix} \quad (6\text{-}26)$$

$$\underline{OO_1}(R_0) = \begin{pmatrix} 0 \\ 0 \\ \ell_0 \end{pmatrix} \quad (6\text{-}27)$$

The definitive solution is:

$$\underline{OO_5}(R_0) = \begin{pmatrix} X_{O_5} \\ Y_{O_5} \\ Z_{O_5} \end{pmatrix} = M_4^0 \begin{pmatrix} 0 \\ \ell_4 \\ 0 \end{pmatrix} + M_3^0 \begin{pmatrix} 0 \\ \ell_3 + TY_3 \\ 0 \end{pmatrix} \quad (6\text{-}28)$$

$$+ M_2^0 \begin{pmatrix} 0 \\ \ell_2 \\ 0 \end{pmatrix} + M_1^0 \begin{pmatrix} 0 \\ 0 \\ \ell_1 \end{pmatrix} + \begin{pmatrix} 0 \\ 0 \\ \ell_0 \end{pmatrix}$$

and is applied to:

$$X_{O_5}(R_0) = S1C(2+4)\ell_4 + S1C2(\ell_3 + TY_3 + \ell_2) \quad (6\text{-}29)$$

$$Y_{O_5}(R_0) = C1C(2+4)\ell_4 + C1C2(\ell_3 + TY_3 + \ell_2) \quad (6\text{-}30)$$

$$Z_{O_5}(R_0) = S(2+4)\ell_4 - S2(\ell_3 + TY_3 + \ell_2) + \ell_1 + \ell_0 \quad (6\text{-}31)$$

The coordinates of O_4 are obtained in the process of calculation ($\ell_4 = 0$ in the preceding equations).

These calculations mean that the position and orientation of the end effector in the fixed coordinate set can be found as a function of articulated variables:

1. The orientation of any vector $\underline{V}$ linked to the end effector can be found:

$$\underline{V}(R_0) = M_4^0 \underline{V}(R_4) \qquad (6\text{-}32)$$

2. The position of a point on the end effector which is known relative to O_5 (in R_4) can be found relative to O (in R_0), since $X_{O_s}(R_0)$, $Y_{O_s}(R_0)$ and $Z_{O_s}(R_0)$ are known.

6.1.3.3 Condensed representation

The above operations show the formation of a geometrical robot model. This model is written in the form of a vector equation, since this allows the notation to be simplified:

$$X(R_0) = F(\underline{\Theta} - \underline{\Theta}_0) \qquad (6\text{-}33)$$

A system of equations is derived:

$$\begin{aligned} x_1 &= f_1(\theta_1 - \theta_{10}, \theta_2 - \theta_{20}, \theta_q - \theta_{q0}) \\ x_2 &= f_2(\theta_1 - \theta_{10}, \theta_2 - \theta_{20}, \theta_q - \theta_{q0}) \\ x_p &= f_p(\theta_1 - \theta_{10}, \theta_2 - \theta_{20}, \theta_q - \theta_{q0}) \end{aligned} \qquad (6\text{-}34)$$

The x_i, which are components of $\underline{X}$, expressed in the coordinate set (R_0), represent the relevant values at the end effector. Using the preceding example these would be:

1. X_{O_s}, Y_{O_s}, Z_{O_s} which define the position of O_5 in (R_0) as a function of the articulated variables θ_i;
2. e_x, e_y, e_z projected along the axes of (R_0), which define the orientation of the end effector (matrix M_4^0).

The articulated variables, also called the *generalized variables*, are the components of vector $\underline{\Theta}$. Vector $\underline{\Theta}_0$ is formed of the articulated variable values, from which the variations are measured. It constitutes their origin.

Continuing with the same example:

$$\underline{\Theta} = (\theta_1, \theta_2, TY_3, \theta_4)^\tau \qquad (6\text{-}35)$$

A vector is represented vertically, and when written horizontally the transposition is shown by the sign τ.

$$\underline{\Theta}_0 = (0, 0, 0, 0)^\tau \qquad (6\text{-}36)$$

For Figure 6.3, the following system of equations can be adopted as the geometrical model:

$$X_{O_s}(R_0) = S1C(2+4)\ell_4 + S1C2(\ell_2 + \ell_3 + TY_3) \tag{6-37}$$
$$Y_{O_s}(R_0) = C1C(2+4)\ell_4 + C1C2(\ell_2 + \ell_3 + TY_3) \tag{6-38}$$
$$Z_{O_s}(R_0) = -S(2+4)\ell_4 - S2(\ell_2 + \ell_3 + TY_3) + \ell_1 + \ell_0 \tag{6-39}$$
$$e_x^4(OX_0) = C1 \tag{6-40}$$
$$e_x^4(OY_0) = S1C(2+4) \tag{6-41}$$
$$e_x^4(OZ_0) = S1S(2+4) \tag{6-42}$$
$$e_y^4(OX_0) = -S1 \tag{6-43}$$
$$e_y^4(OY_0) = C1C(2+4) \tag{6-44}$$
$$e_y^4(OZ_0) = C1S(2+4) \tag{6-45}$$
$$e_z^4(OX_0) = 0 \tag{6-46}$$
$$e_z^4(OY_0) = -S(2+4) \tag{6-47}$$
$$e_z^4(OZ_0) = C(2+4) \tag{6-48}$$

6.1.3.4 Redundant equations

First type of redundancy: The nine equations (6-40) to (6-48) represent the projections of the three unity vectors of the coordinate set (R_4) along the axes of (R_0): $\underline{e_x^4}$, $\underline{e_y^4}$, $\underline{e_z^4}$. These are not independent, and the following relationships exist:

$$|\underline{e_x^4}| = |\underline{e_y^4}| = |\underline{e_z^4}| = 1 \tag{6-49}$$
$$\underline{e_x^4} \cdot \underline{e_y^4} = \underline{e_x^4} \cdot \underline{e_z^4} = \underline{e_y^4} \cdot \underline{e_z^4} = 0 \tag{6-50}$$

Applying these to equations (6-40) to (6-48):

$$[e_x^4(OX_0)]^2 + [e_x^4(OY_0)]^2 + [e_x^4(OZ_0)]^2 = 1 \tag{6-51}$$
$$[e_y^4(OX_0)]^2 + [e_y^4(OY_0)]^2 + [e_y^4(OZ_0)]^2 = 1 \tag{6-52}$$
$$[e_z^4(OX_0)]^2 + [e_z^4(OY_0)]^2 + [e_z^4(OZ_0)]^2 = 1 \tag{6-53}$$
$$e_x^4(OX_0)e_y^4(OX_0) + e_x^4(OY_0)e_y^4(OY_0) + e_x^4(OZ_0)e_y^4(OZ_0) = 0 \tag{6-54}$$
$$e_x^4(OX_0)e_z^4(OX_0) + e_x^4(OY_0)e_z^4(OY_0) + e_x^4(OZ_0)e_z^4(OZ_0) = 0 \tag{6-55}$$
$$e_y^4(OX_0)e_z^4(OX_0) + e_y^4(OY_0)e_z^4(OY_0) + e_y^4(OZ_0)e_z^4(OZ_0) = 0 \tag{6-56}$$

For the nine equations (6-40) to (6-48) no more than three interdependent relationships can be found. These are given by:

- any component of $\underline{e}_x^4$, $\underline{e}_y^4$, $\underline{e}_z^4$;
- or any two components of an $\underline{e}_i^4$ and one of an $\underline{e}_j^4$ ($j \neq i$).

Three equations are chosen for the calculations necessary for robot control.

Second type of redundancy: If only three of the preceding equations, for example $e_x(OX_0)$, $e_y(OY_0)$, $e_z(OZ_0)$, are taken, along with the three essential equations giving X_{O_s}, Y_{O_s}, Z_{O_s}, the model consists of six equations. There are, however, four variables: θ_1, θ_2, TY_3, θ_4. This second type of redundancy is discussed in the context of control.

6.2 Geometrical or positional control

The geometrical model is expressed in equation (6-33). Its direct application is that if the value of each articulation, and, of course, the length of each segment are known, the position and orientation of the end effector (represented by the position of a point, and the orientation of a trihedron linked to the end effector) in real space — the task-space being represented by coordinate set (R_0) — can be found.

To control the robot, equation (6-33) must be made to work in reverse. If the required position and orientation of the end effector in real space are known, then the required values of the articulated variables (of $\underline{\Theta}$) must be found. These values can be found by inverting equation (6-33):

$$\underline{\Theta} - \underline{\Theta}_0 = F^{-1}[\underline{X}(R_0)] \qquad (6\text{-}57)$$

that is:

$$\begin{aligned} \theta_1 - \theta_{10} &= g_1(x_1, x_2, x_p) \\ \theta_2 - \theta_{20} &= g_2(x_1, x_2, x_p) \\ \theta_q - \theta_{q0} &= g_q(x_1, x_2, x_p) \end{aligned} \qquad (6\text{-}58)$$

If equation (6-57) can be solved for a given robot, it is said to be *resolvable*, and only in this circumstance can geometrical control be established. Equation (6-57) is a *positional control algorithm*.

6.2.1 PROBLEMS ASSOCIATED WITH ROBOT RESOLVABILITY

There are several problems attached to the robot with four DOF, as shown in Figure 6.3, with which the following geometrical model is associated:

$$X_{O_5}(R_0) = S1C(2+4)\ell_4 + S1C2(\ell_2 + \ell_3 + TY_3)$$
$$Y_{O_5}(R_0) = C1C(2+4)\ell_4 + C1C2(\ell_2 + \ell_3 + TY_3)$$
$$Z_{O_5}(R_0) = S(2+4)\ell_4 - S2(\ell_2 + \ell_3 + TY_3) + \ell_1 + \ell_0$$
$$e_y^4(OX_0) = -S1$$
$$e_y^4(OY_0) = C1C(2+4)$$
$$e_z^4(OZ_0) = C(2+4)$$
(6-59)

If the following information is known:

- the structure, ie the lengths ℓ_0, ℓ_1, ℓ_2, ℓ_3, ℓ_4 and the values θ_{10}, θ_{20}, TY_{30}, θ_{40} from which the articulated variables are measured (reference values are nil);
- the position of O_5 in the fixed coordinate set X_{O_5}, Y_{O_5}, Z_{O_5};
- the orientation of $O_4 O_5$ relative to (R_0) (for which there are many methods of representation, but in this example the cosine projectors $e_x^4(OX_0)$, $e_y^4(OY_0)$ and $e_z^4(OZ_0)$ are chosen).

How can the values of θ_1, θ_2, TY_3 and θ_4 be found?

The first problem arises from considerations of elementary geometry. The robot is too far from the required point O_5 (see Figure 6.4). This point is not physically attainable. Equation (6-59) offers no real solution.

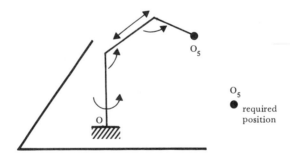

Figure 6.4. *Robot remote from required position O_5*

The second problem arises from considerations of elementary mechanics. It is assumed, when dealing with robot geometry, that each rotary articulation can turn through 360 degrees, and that each translation is infinite (in which case the problem of non-attainable points only occurs in robots with entirely rotational articulations). In practice, translations have limited amplitude, as do rotations. Consequently, the equations can provide solutions that are not mechanically attainable.

The third problem states although it is known that there are geometrical and mechanical solutions to the problems, equation (6-59) is made up of six non-linear equations, with four unknowns. This is a mathematical problem.

Using the previous example, several solutions are possible:
1. It is known that $X_{O_5}(R_0) = A$, $Y_{O_5}(R_0) = B$, $Z_{O_5}(R_0) = C$. O_5 is to be positioned, without taking into account the orientation of $O_4 O_5$. The following system of equations applies:

$$A = S1[C(2+4)\ell_4 + C2(\ell_2 + \ell_3 + TY_3)] \quad (6\text{-}60)$$

$$B = C1[C(2+4)\ell_4 + C1C2(\ell_2 + \ell_3 + TY_3)] \quad (6\text{-}61)$$

$$C = -S(2+4)\ell_4 - S2(\ell_2 + \ell_3 + TY_3) + \ell_1 + \ell_0 \quad (6\text{-}62)$$

(a) If the solutions to the problem are physically realizable, with three equations and four unknowns, the possible solutions are infinite (given by infinite combinations of θ_2, TY_3 and θ_4 since θ_1 is defined as π using $\tan\theta_1 = A/B$).

(b) Since the equations are non-linear with respect to the variable, the analytical solutions are difficult to find.

(c) To find a finite number of solutions there must be a choice criterion, such as leaving one articulation in a fixed position.

2. If, in addition, the orientation of $O_4 O_5$ (represented by e_y^4) is to be found in the coordinate set (R_0) then:

$$e_y^4(OX_0) = \alpha = -S1 \quad (6\text{-}43)$$

$$e_y^4(OY_0) = \beta = C1C(2+4) \quad (6\text{-}44)$$

$$e_y^4(OZ_0) = \gamma = C1S(2+4) \quad (6\text{-}45)$$

The three equations are not independent, since $|e_y^4| = 1$. Two are sufficient, for example the first two. But since θ_1 is fixed (at a value of approximately π) when point O_5 is fixed (which means that the direction of $O_4 O_5$ cannot be chosen arbitrarily once the position of O_5 is chosen) one equation which does not only depend on θ_1 is sufficient. Thus:

$$\beta = C1C(2+4) \quad (6\text{-}63)$$

Equations (6-60) to (6-62) and (6-44) make up a system of four compatible equations with four unknowns.

A robot is said to be *resolvable* if from a number of compatible equations equal to the number of unknowns, the values of the unknowns can be found analytically and the correct solution selected (non-linearity provides a number of solutions).

If the system in equations (6-60) to (6-63) is to be resolved, it should be noted that equations (6-60) and (6-61) allow $\tan\theta_1 = A/B$ to be found, from which two values of θ_1 can be determined: $\theta_{11} = \theta_1^x$ and $\theta_{12} = \theta_1^x - \pi$. (This involves two opposing values for the sine and cosine: $\theta_{11} \leftarrow C\theta_1^x S\theta_1^x$; $\theta_{12} \rightarrow -C\theta_1^x - S\theta_1^x$).

Equation (6-63) provides two opposing values for C(2+4), which leads to four possibilities for $\theta_2 + \theta_4$: $\theta_2^x + \theta_4^x$; $\pi - (\theta_2^x + \theta_4^x)$; $\pi + (\theta_2^x + \theta_4^x)$; $-(\theta_2^x + \theta_4^x)$ and two possible values for S(2+4). Equations (6-60) and (6-62) thus can give, for example:

$$C2(\ell_2 + \ell_3 + TY_3) = \frac{A}{S1} - C(2+4)\ell_4 \qquad (6\text{-}64)$$

$$S2(\ell_2 + \ell_4 + TY_3) = \ell_1 + \ell_0 - C - S(2+4)\ell_4 \qquad (6\text{-}65)$$

and:

$$\tan \theta_2 = \frac{\ell_1 + \ell_0 - C - \ell_4 S(2+4)}{\frac{A}{S1} - \ell_4 C(2+4)} \qquad (6\text{-}66)$$

$$TY_3 = -(\ell_2 + \ell_3) + \frac{\ell_1 + \ell_0 - C - \ell_4 S(2+4)}{S2} \qquad (6\text{-}67)$$

This gives rise to four possible values for $\tan \theta_2$, and eight values for θ_2. TY_3 could assume one of four values. For θ_4, eight values can be derived from the sum of $\theta_2 + \theta_4$.

These eight solutions are geometrical, and do not take into account the issue of physical non-attainability caused by articulation limits. Figure 6.5 summarizes the various possibilities.

To eliminate the unacceptable solutions, consider that a translation may not be negative. Thus, four solutions are eliminated.

If θ_1 can turn through 360 degrees, θ_2 cannot be made to turn mechanically through more than π degrees (from $-\pi/2$ to $+\pi/2$, since in Figure 6.3 the robot was represented with $\theta_2 = 0$).

Thus, the physical solutions are reduced to:

$$\theta_1^x, \theta_{21}^x, TY_3(\theta_{21}^x), \theta_{41}^x = \theta_2^x + \theta_4^x - \theta_{21}^x \qquad (6\text{-}68)$$

$$\theta_1^x, \theta_{22}^x, TY_3(\theta_{22}^x), \theta_{42}^x = -(\theta_2^x + \theta_4^x) - \theta_{22}^x \qquad (6\text{-}69)$$

If O_5 and the direction of $O_4 O_5$ are imposed, O_4 is imposed in position. If the translation is positive, there is only one solution for θ_1, θ_2 and TY_3. In fact, if numerical values are applied to the two remaining solutions, one will always lead to a non-attainable configuration.

In systems with more than four DOF, there are often several mechanical solutions, and the choice must be made by imposing additional constraints.

6.2.2 ADVANTAGES AND DISADVANTAGES OF POSITIONAL CONTROL

If the robot shown in Figure 6.3 is required to position the point O_5 relative to the coordinate set (R_0), the following facts are known:

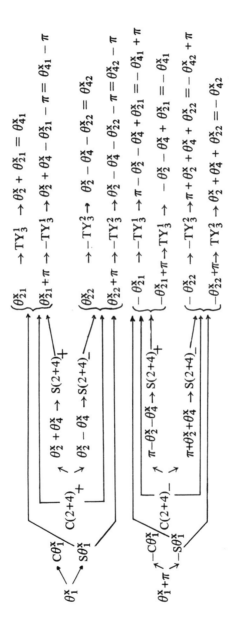

Figure 6.5. *All possible solutions to the system in equations (6-60) to (6-63)*

1. It must calculate the values of the articulations as explained above (not all robots are resolvable).
2. It must measure the actual values of its articulated variables.
3. It must translate the difference between actual and required values into power signals fed to the motors driving the articulations. For positional control the servocontrol motors are made to work until the articulation sensors (internal) indicate that the required position has been reached, and the motor should stop.

This causes many problems:

1. The trajectory followed by point O_5 depends on the level of excitation of the motors (or their parallel control).
2. The speed of movement only can be controlled by supplying the motors at a higher or lower level.
3. Since inertia and friction are not taken into account, there is a risk of oscillation and of overshooting the final position if movements are made at high speed.

In geometrical control the assumption is made that balanced successive configurations are passed through, moving from rest each time. This is why large movements are divided into successive smaller trajectories, for which the model is more valid. This type of control is widely used in industrial robots.

Chapter 7
Control based on the kinematic model

7.1 Kinematic model: a simplified robot model

It has been shown that the geometrical model cannot always be used, for example when a robot is non-resolvable, and gives an approximation that is not encountered in practice. This sometimes results in control problems (speed control, accuracy and overshooting etc).

Another type of model, with various improvements, can be established. This model is known by at least three names: the *kinematic model*, the *variation model* and the *speed control model*. It consists of replacing the geometrical model, in which the main problem is the non-linearity in relation to the articulated variables, with a linear approximation which is valid if small movements are made. Thus, the model $\underline{X} = \underline{F}(\Theta - \Theta_0)$ (equation 6-33) is replaced by its increments:

$$\Delta \underline{X} = \frac{\partial \underline{F}}{\partial \underline{\Theta}} \Delta \underline{\Theta} \tag{7-1}$$

which makes the variations of $\underline{X}$ linear as a function of Θ. Thus the equation (6-34):

$$\begin{aligned} x_1 &= f_1(\theta_1 - \theta_{10}, \theta_2 - \theta_{20}, \theta_q - \theta_{qo}) \\ x_2 &= f_2(\theta_1 - \theta_{10}, \theta_2 - \theta_{20}, \theta_q - \theta_{qo}) \\ x_p &= f_p(\theta_1 - \theta_{10}, \theta_2 - \theta_{20}, \theta_q - \theta_{qo}) \end{aligned} \tag{6-34}$$

is replaced by:

$$\begin{aligned} \Delta x_1 &= \frac{\partial f_1}{\partial \theta_1}\Delta\theta_1 + \frac{\partial f_1}{\partial \theta_2}\Delta\theta_2 + \frac{\partial f_1}{\partial \theta_q}\Delta\theta_q \\ \Delta x_2 &= \frac{\partial f_2}{\partial \theta_1}\Delta\theta_1 + \frac{\partial f_2}{\partial \theta_2}\Delta\theta_2 + \frac{\partial f_2}{\partial \theta_q}\Delta\theta_q \\ \Delta x_p &= \frac{\partial f_p}{\partial \theta_1}\Delta\theta_1 + \frac{\partial f_p}{\partial \theta_2}\Delta\theta_2 + \frac{\partial f_p}{\partial \theta_q}\Delta\theta_q \end{aligned} \tag{7-2}$$

Equation (7-2) is linear and can be written in matrix form:

$$\begin{pmatrix} \Delta x_1 \\ \Delta x_2 \\ \Delta x_p \end{pmatrix} = \begin{pmatrix} \frac{\partial f_1}{\partial \theta_1} & \frac{\partial f_1}{\partial \theta_2} & \frac{\partial f_1}{\partial \theta_q} \\ \frac{\partial f_2}{\partial \theta_1} & \frac{\partial f_2}{\partial \theta_2} & \frac{\partial f_2}{\partial \theta_q} \\ \frac{\partial f_p}{\partial \theta_1} & \frac{\partial f_p}{\partial \theta_2} & \frac{\partial f_p}{\partial \theta_q} \end{pmatrix} \begin{pmatrix} \Delta \theta_1 \\ \Delta \theta_2 \\ \Delta \theta_q \end{pmatrix} \quad (7\text{-}3)$$

Equation (7-3) is the developed expression of equation (7-1). The matrix:

$$[J] = \frac{\partial \underline{F}}{\partial \underline{\Theta}} \quad (7\text{-}4)$$

is called the Jacobian of F or the Jacobian matrix. It can also be expressed thus:

$$\underline{\Delta X} = [J] \underline{\Delta \Theta} \quad (7\text{-}5)$$

This is the *kinematic model*.

7.1.1 EXAMPLE

Consider the geometrical model given in equations (6-37) to (6-48). The first equation is:

$$X_{O_s}(R_0) = S1C(2+4)\ell_4 + S1C2(\ell_2 + \ell_3 + TY_3) \quad (6\text{-}37)$$

from which:

$$\Delta X_{O_s}(R_0) = \frac{\partial f}{\partial \theta_1}\Delta\theta_1 + \frac{\partial f}{\partial \theta_2}\Delta\theta_2 + \frac{\partial f}{\partial TY_3}\Delta TY_3 + \frac{\partial f}{\partial \theta_4}\Delta\theta_4 \quad (7\text{-}6)$$

$$\frac{\partial f}{\partial \theta_1} = C1C(2+4)\ell_4 + C1C2(\ell_2 + \ell_3 + TY_3) \quad (7\text{-}7)$$

$$\frac{\partial f}{\partial \theta_2} = -S1S(2+4)\ell_4 - S1S2(\ell_2 + \ell_3 + TY_3) \quad (7\text{-}8)$$

$$\frac{\partial f}{\partial TY_3} = S1C2 \quad (7\text{-}9)$$

$$\frac{\partial f}{\partial \theta_4} = -S1S(2+4)\ell_4 \quad (7\text{-}10)$$

In the same way:

$$\Delta Y_{O_s}(R_0) = [-S1[C(2+4)\ell_4 + C_2(\ell_2 + \ell_3 + TY_3)]]\Delta\theta_1$$
$$-C1[S(2+4)\ell_4 + S2(\ell_2 + \ell_3 + TY_3)]\Delta\theta_2$$
$$+ C1C2\Delta TY_3 - C1S(2+4)\ell_4 \Delta\theta_4 \quad (7\text{-}11)$$

$$\Delta Z_{O_s}(R_0) = [-C(2+4)\ell_4 - C2(\ell_2 + \ell_3 + TY_3)]\Delta\theta_2$$
$$- S2\Delta TY_3 - C(2+4)\ell_4 \Delta\theta_4 \quad (7\text{-}12)$$

$$\Delta e_x^4(OX_0) = S1\Delta\theta_1 \quad (7\text{-}13)$$

$$\Delta e_x^4(OY_0) = C1C(2+4)\Delta\theta_1 - S1S(2+4)\Delta\theta_2 - S1S(2+4)\Delta\theta_4 \quad (7\text{-}14)$$

$$\Delta e_x^4(OZ_0) = C1S(2+4)\Delta\theta_1 + S1C(2+4)\Delta\theta_2 + S1C(2+4)\Delta\theta_4 \quad (7\text{-}15)$$

$$\Delta e_y^4(OX_0) = -C1\Delta\theta_1 \quad (7\text{-}16)$$

$$\Delta e_y^4(OY_0) = -S1C(2+4)\Delta\theta_1 - C1S(2+4)\Delta\theta_2 - C1S(2+4)\Delta\theta_4 \quad (7\text{-}17)$$

$$\Delta e_y^4(OZ_0) = -S1S(2+4)\Delta\theta_1 + C1C(2+4)\Delta\theta_2 + C1C(2+4)\Delta\theta_4 \quad (7\text{-}18)$$

$$\Delta e_z^4(OX_0) = 0 \quad (7\text{-}19)$$

$$\Delta e_z^4(OY_0) = -C(2+4)\Delta\theta_2 - C(2+4)\Delta\theta_4 \quad (7\text{-}20)$$

$$\Delta e_z^4(OZ_0) = -S(2+4)\Delta\theta_2 - S(2+4)\Delta\theta_4 \quad (7\text{-}21)$$

Thus, the variational model can be written in matrix form:

$$\begin{vmatrix} \Delta X_{O_s}(R_0) \\ \Delta Y_{O_s}(R_0) \\ \Delta Z_{O_s}(R_0) \\ \Delta e_x^4(OX_0) \\ \\ \\ \\ \\ \\ \\ \\ \Delta e_z^4(OZ_0) \end{vmatrix} \quad \begin{vmatrix} \alpha_1 & \beta_1 & \gamma_1 & \delta_1 \\ \alpha_2 & \beta_2 & \gamma_2 & \delta_2 \\ 0 & \beta_3 & \gamma_3 & \delta_3 \\ \alpha_4 & 0 & 0 & 0 \\ \alpha_5 & \beta_5 & 0 & \beta_5 \\ \alpha_6 & \beta_6 & 0 & \beta_6 \\ \alpha_7 & 0 & 0 & 0 \\ -\beta_6 & -\alpha_6 & 0 & -\alpha_6 \\ \beta_5 & \alpha_5 & 0 & \alpha_5 \\ 0 & 0 & 0 & 0 \\ 0 & \beta_7 & 0 & \beta_7 \\ 0 & \beta_8 & 0 & \beta_8 \end{vmatrix} \quad \begin{vmatrix} \Delta\theta_1 \\ \Delta\theta_2 \\ \Delta TY_3 \\ \Delta\theta_4 \end{vmatrix} \quad (7\text{-}22)$$

$$12 \times 1 \qquad\qquad 12 \times 4 \qquad\qquad 4 \times 1$$

In this example the Jacobian J has 12 rows and four columns. In the nine equations concerning cosine directrix, six, in principle, can be deleted (so long as the remaining three are independent). The resulting Jacobian still corresponds to the same problem, but only has six rows and four columns. In fact, as shown in Section 6.2.1, this could be reduced to four equations with four unknowns in a specific case (if O_5 and the direction $O_4 O_5$ are imposed).

7.1.2 VARIATIONAL MODEL

Starting from a configuration in which $\underline{X}$ and $\underline{\Theta}$ are known, the equation:

$$\underline{\Delta X} = [J(\Theta)] \underline{\Delta \Theta} \qquad (7\text{-}5)$$

allows $\underline{\Delta X}$ to be found, and thus the new position of $\underline{X} + \underline{\Delta X}$, if $\underline{\Delta \Theta}$ is imposed from $\underline{\Theta}$.

7.2 Variational control

As has been stated already, it is as important to find $\underline{\Theta}$ when $\underline{X}$ is known as to find $\underline{X}$ when $\underline{\Theta}$ is known. Both are necessary for robot control.

In the kinematic model of equation (7-5) which is linear with respect to $\underline{\Delta \Theta}$, inversion is easier than for the geometrical model. Thus:

$$\underline{\Delta X} = [J(\Theta)] \underline{\Delta \Theta} \qquad (7\text{-}5)$$

$$\underline{\Delta \Theta} = [J(\Theta)]^{-1} \underline{\Delta X} \qquad (7\text{-}23)$$

Thus, starting from a configuration in which $\underline{\Theta}$ and $\underline{X}$ are known, if $\underline{\Delta X}$ is to be found, equation (7-23) allows $\underline{\Delta \Theta}$ to be found by inverting the matrix $[J]$, an operation explained in Appendix I. This inversion is sometimes less simple than it seems, in particular cases.

7.2.1 IF [J] IS NOT SQUARE

Inversion cannot be carried out using standard methods. The theory of generalized inverses (not explained here) must be used.

7.2.2 IF [J] IS SQUARE AND ITS DETERMINANT NOT EQUAL TO ZERO

Inversion is possible, as shown in this simple example: point O_4 of the robot in Figure 6.3 is to be controlled.

$$X_{O_4}(R_0) = S1C2(L_2 + TY_3) \qquad (7\text{-}24)$$

$$Y_{O_4}(R_0) = C1C2(L_2 + TY_3) \qquad (7\text{-}25)$$

$$Z_{O_4}(R_0) = -S2(L_2 + TY_3) + L_1 \qquad (7\text{-}26)$$

To simplify these equations, the following notation will be adopted: $L_2 = \ell_2 + \ell_3$; $L_1 = \ell_0 + \ell_1$. The equations can be differentiated:

$$\Delta X_{O_4}(R_0) = [C1C2(L_2 + TY_3)]\Delta\theta_1 - [S1S2(L_2 + TY_3)]\Delta\theta_2$$
$$+ S1C2\Delta TY_3 \qquad (7\text{-}27)$$

$$\Delta Y_{O_4}(R_0) = [-S1C2(L_2 + TY_3)]\Delta\theta_1 - [C1S2(L_2 + TY_3)]\Delta\theta_2$$
$$+ C1C2\Delta TY_3 \qquad (7\text{-}28)$$

$$\Delta Z_{O_4}(R_0) = -C2(L_2 + TY_3)\Delta\theta_2 - S2\Delta TY_3 \qquad (7\text{-}29)$$

Matrix [J] is derived:

$$[J] = \begin{pmatrix} C1C2(L_2 + TY_3) & -S1S2(L_2 + TY_3) & S1C2 \\ -S1C2(L_2 + TY_3) & -C1S2(L_2 + TY_3) & C1C2 \\ 0 & -C2(L_2 + TY_3) & -S2 \end{pmatrix} \quad (7\text{-}30)$$

If [J] is inverted, the result is:

$$[J]^{-1} = \frac{1}{C2(L_2 + TY_3)}$$

$$\times \begin{vmatrix} C1 & -S1 & 0 \\ -S1S2C2 & -C1C2S2 & -C^22 \\ S1C^22(L_2 + TY_3) & C1C^22(L_2 + TY_3) & -S2C^22(L_2 + TY_3) \end{vmatrix} \quad (7\text{-}31)$$

The following system of equations is found:

$$\Delta\theta_1(L_2 + TY_3) = (C1/C2)\Delta X_{O_4} - (S1/C2)\Delta Y_{O_4} \qquad (7\text{-}32)$$
$$\Delta\theta_2(L_2 + TY_3) = -S1S2\Delta X_{O_4} - C1S2\Delta Y_{O_4} - C2\Delta Z_{O_4} \qquad (7\text{-}33)$$
$$\Delta TY_3 = S1C2\Delta X_{O_4} + C1C2\Delta Y_{O_4} - S2\Delta Z_{O_4} \qquad (7\text{-}34)$$

7.2.3 IF [J] IS SQUARE AND ITS DETERMINANT EQUAL TO ZERO FOR CERTAIN VALUES OF THE ARTICULATED VARIABLES

If only the geometry of the system (rotations of 360 degrees possible) is taken into account, then this is always the case. Thus, in the preceding example $\Delta = C2(L_2 + TY_3) = 0$ for $\theta_2 = \pi/2$, $\theta_2 = -\pi/2$ or for $TY_3 = -L_2$. This clearly corresponds (see Figure 7.1) to either (a) a negative translation (O_4 confused with O_2) or (b) having $O_2 O_3$ on the axis OZ_0.

In practice, it is best to arrange that the determinant should not equal zero, when calculating articulation limits. These configurations exist, however, in some robots. These are called *singular configurations* or *robot singularities*.

7.2.4 PHYSICAL MEANING OF A SINGULARITY

Taking as an example equations (7-27) to (7-29) in which $\theta_2 = +\pi/2$,

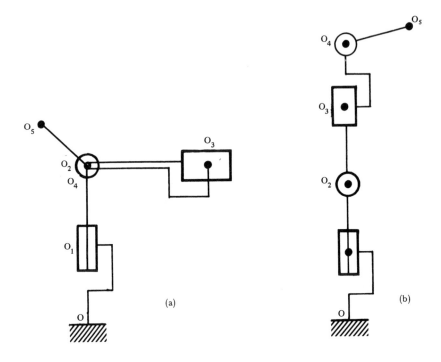

Figure 7.1. *Examples of singular configurations: (a) $TY_3 = -L_2$; (b) $\theta_2 = \pm\pi/2$*

the following is obtained:

$$\Delta X_{O_4}(R_0) = -S1(L_2 + TY_3)\Delta\theta_2 \qquad (7\text{-}35)$$

$$\Delta Y_{O_4}(R_0) = -C1(L_2 + TY_3)\Delta\theta_2 \qquad (7\text{-}36)$$

$$\Delta Z_{O_4}(R_0) = -\Delta TY_3 \qquad (7\text{-}37)$$

(a) It can be seen that $\Delta\theta_1$ does not occur in the equations, thus it cannot be calculated. In Figure 7.1 O, O_1, O_2, O_3 and O_4 are aligned. Point O_4 has the coordinates O, O, $L_1 + L_2 + TY_3$. Nothing can be determined in advance for the value of θ_1 to satisfy equations (7-35) to (7-37).

(b) Otherwise the equations provide $\Delta X_{O_4}/\Delta Y_{O_4} = \tan\theta_1$. ΔX_{O_4}, ΔY_{O_4} and ΔZ_{O_4} cannot be chosen arbitrarily.

Thus, a singularity corresponds to an indeterminate value of a variable. This corresponds in a matrix J to a lowering of its rank. (The rank is equal to the dimension of the largest non nil determinant in the matrix.)

So with the example, if $\theta_2 = \pi/2$:

$$[J] = \begin{pmatrix} 0 & -S1(L_2 + TY_3) & 0 \\ 0 & -C1(L_2 + TY_3) & 0 \\ 0 & 0 & -1 \end{pmatrix} \qquad (7\text{-}38)$$

The larger of the two non nil determinants is for example:

$$|J_1| = \begin{vmatrix} -S1(L_2 + TY_2) & 0 \\ 0 & -1 \end{vmatrix} \quad (7\text{-}39)$$

$[J_1]$ is of rank 2 (if $\theta_1 \neq K\pi$ and $L_2 \neq TY_3$). This singularity is said to be of the first degree since J is of dimension 3, and is the larger non nil determinant of dimension 2. Note: if in addition to $\theta_2 = \pi/2$, $L_2 = -TY_3$, then the largest non nil determinant would be:

$$|J_2| = |-1| \quad (10\text{-}40)$$

The singularity would be of the second degree:

$$[= (\text{dimension of J}) - (\text{dimension of } J_2)]$$

The first example, concerning first-degree singularity will be considered.

7.2.5 TO RESOLVE THE EQUATION SYSTEM IN THE EVENT OF A SINGULARITY

There are various methods for this. The simplest is the principal variable method: [J] for $\theta_2 = \pi/2$ is as follows:

$$[J] = \begin{pmatrix} \Delta\theta_1 & \Delta\theta_2 & \Delta TY_3 \\ 0 & -S1(L_2 + TY_3) & 0 \\ 0 & -C1(L_2 + TY_3) & 0 \\ 0 & 0 & -1 \end{pmatrix} \quad (7\text{-}41)$$

The first column corresponds to the terms multiplied by $\Delta\theta_1$, the second to terms multiplied by $\Delta\theta_2$, the third to terms multiplied by ΔTY_3. Since the first column is nil, a singularity arises from $\Delta\theta_1$. The other columns are not nil (assuming that $L_2 \neq -TY_3$), and so there are no other singularities. The method of principal variables is reasoned as follows:

1. Since $\Delta\theta_1$ cannot be calculated, and the equations do not depend on it, $\Delta\theta_1 = 0$ will be chosen.
2. The problem is resolved with $\Delta\theta_2$ and ΔTY_3, which are the principal variables.
3. The system of equations is thus defined by equations (7-35) to (7-37) which give the result:

$$\Delta\theta_2 = -\Delta X_{O_4}/S1(L_2 + TY_3) = -\Delta Y_{O_4}/C1(L_2 + TY_3) \quad (7\text{-}42)$$
$$\Delta TY_3 = -\Delta Z_{O_4} \quad (10\text{-}43)$$

with:

$$\Delta\theta_1 = 0 \quad (7\text{-}44)$$

It should be noted that, in this example, although it is not always so, the singularity causes a constraint between ΔX_{O_4} and ΔY_{O_4}, which cannot be chosen independently of one another.

In matrix calculation, a sub matrix with the largest possible dimension, and non nil determinant, should be considered for $[J]$. One of the systems shown below can be chosen:

$$\begin{pmatrix} \Delta X_{O_4} \\ \Delta Z_{O_4} \\ \Delta Y_{O_4} \end{pmatrix} = \begin{pmatrix} 0 & -S1(L_2 + TY_3) & 0 \\ 0 & 0 & -1 \\ 0 & -C1(L_2 + TY_3) & 0 \end{pmatrix} \begin{pmatrix} \Delta\theta_1 \\ \Delta\theta_2 \\ \Delta TY_3 \end{pmatrix} \quad (7\text{-}45)$$

or:

$$\begin{pmatrix} \Delta X_{O_4} \\ \Delta Y_{O_4} \\ \Delta Z_{O_4} \end{pmatrix} = \begin{pmatrix} 0 & -S1(L_2 + TY_3) & 0 \\ 0 & 0 & -1 \\ 0 & -C1(L_2 + TY_3) & 0 \end{pmatrix} \begin{pmatrix} \Delta\theta_1 \\ \Delta\theta_2 \\ \Delta TY_3 \end{pmatrix} \quad (7\text{-}46)$$

7.2.6 IF $[J]$ IS SQUARE WITH DETERMINANT EQUAL TO ZERO, WHATEVER THE VARIABLE VALUES

This indicates that the equations which allow $[J]$ to be formed are not independent. If the problem of orienting the axis $O_4 X_4$ in the space (R_0)(see Figure 6.3) is considered, the variations of the cosine projectors $e_x^4(OX_0)$, $e_y^4(OY_0)$ and $e_z^4(OZ_0)$ are given by:

$$\Delta e_x^4(OX_0) = -S1\Delta\theta_1 \quad (7\text{-}13)$$

$$\Delta e_x^4(OY_0) = C1C(2+4)\Delta\theta_1 - S1S(2+4)\Delta\theta_2 - S1S(2+4)\Delta\theta_4 \quad (7\text{-}14)$$

$$\Delta e_x^4(OZ_0) = C1S(2+4)\Delta\theta_1 + S1C(2+4)\Delta\theta_2 + S1C(2+4)\Delta\theta_4 \quad (7\text{-}15)$$

thus:

$$[J] = \begin{pmatrix} -S1 & 0 & 0 \\ C1C(2+4) & -S1S(2+4) & -S1S(2+4) \\ C1S(2+4) & S1C(2+4) & S1C(2+4) \end{pmatrix} \quad (7\text{-}47)$$

The determinant of $[J]$ is nil, the two columns are identical. This is normal since $|e_y^4| = 1$, and the preceding equations are not independent. This means that the choice of values to be controlled is not correct.

7.3 Characteristics of kinematic control

Kinematic control now can be established, that is the small successive increments of the motor angles which correspond to the small increments in position and orientation of the robot's end effector can be calculated. The velocity and movement can be controlled more fully, since each increment $\Delta\theta_i$ occurs during time T, which gives an articulated velocity of $\dot{\theta}_i = \Delta\theta_i/T$ with T being the control period. The velocity can be varied either by acting on T (but it is more practical to maintain this constant), or on the amplitude of $\Delta\theta_i$ (which is the more common solution). $\Delta\theta_i$ must be small if the model is to remain valid.

As with the geometrical model, the masses, inertia and friction that alter the behaviour of the robot are not taken into consideration, and thus the kinematic model still fails to provide a completely valid solution for control planning. This type of control is used particularly in the control of non-resolvable robots.

7.4 Models and dynamic control

The robot computer uses models to predict the robot's behaviour. An important consideration is that models are only an approximation of reality, and thus only valid under certain conditions.

The models analysed in the previous chapters depend on three assumptions:

1. The robot has a perfect structure: it does not flex under load, the rotational centres of the articulations do not move relative to the structure. The geometry is assumed to correspond with that described in the equations.
2. Control is instantaneous and accurate, and even the smallest movement can be made.
3. There are no dynamic phenomena affecting movement but in practice, this is not true. At high speed, it is difficult to stop instantaneously because of inertial effects, movement cannot be started without friction and vibrations sometimes occur due to the elasticity of the segments and transmissions etc.

For these reasons improved models have been proposed, which offer better control than the geometrical and kinematic models. Further information on this subject can be found in Appendix V.

Chapter 8
Actuator servocontrol

In this chapter the basic principles of servo-systems are discussed. Servo-systems are electro-mechanical systems in which the output, whether in the form of position, speed or torque, is constantly servoed by an input command. In this chapter the problems involved in employing servo-systems in robots will also be considered.

8.1 Principles of servocontrol

8.1.1 BASIC PRINCIPLES

For a servo-motor (see Chapter 5), there is a linear relationship between the input voltage V_I and the speed of rotation ω_O of the output shaft, as shown in Figure 8.1. This linearity depends on the characteristics (ie inertia j and friction f) of the load C.

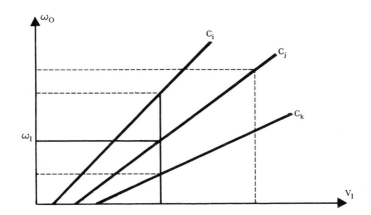

Figure 8.1. *Linear relationship between input voltage V_I and speed of rotation ω_O of the output shaft in the servo-motor*

Thus, to move a given load C at speed ω_C, the servo-motor must be supplied with voltage V_I shown in the curve $\omega_O(V_I)$ to correspond to C (see Figures 8.2a and b).

There is, of course, an interval during which the motor reaches the required speed, and this is known as the *response time* (see Figure 8.2c). If load C is modified (particularly as a result of external disturbance), the output speed for the same input voltage V_I will be different (eg ω_O decreases if friction increases) (see Figure 8.2d).

Thus, for the same input, the output signal will not necessarily be the same. The control of this type of system is said to be *direct* or in *open-loop mode*.

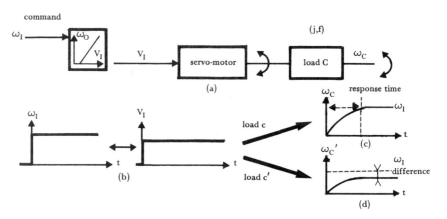

Figure 8.2. *(a) and (b) If the motor is supplied with voltage V_I, shown by curve $\omega_O(V_I)$, a given load C can be handled; (c) response time; (d) if load C is altered the output speed for the same input voltage V_I will alter*

One of the main characteristics of this type of control is that it makes no compensation for the disturbance acting on the system. Since disturbance is, by definition, unpredictable, the only way to combat disturbance is to control the output signal by:

1. measuring the difference between the required speed ω_I (or control) and the speed obtained ω_C;
2. modify the input voltage V_I (to make up for this difference in speed) by an amount ΔV proportional to the difference $\omega_C - \omega_I$. Voltage V_I is then modified until the two speeds are equal ($\Delta V = 0$).

To achieve this in practice, various devices can be used:

1. a *sensor* capable of reading the information at the output shaft, and then converting it into a proportional voltage (eg using a tachometric generator);
2. a device that continuously calculates the difference, known as a *differential amplifier*;
3. a circuit as shown in Figure 8.3.

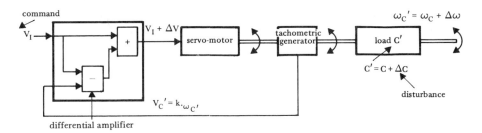

Figure 8.3. *A variation in speed brings about a variation in input control due to feedback*

This system shows how variation in speed causes a variation ΔV in input control (of the opposite sign), which will oppose and counterbalance the effect of the disturbance. This can only be achieved after several variations about the *set point* (see Figure 8.4).

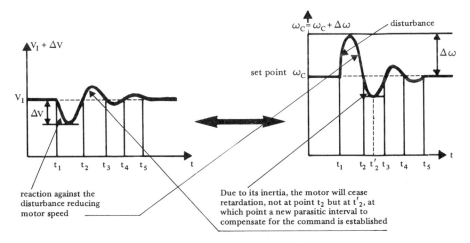

Figure 8.4. *Variations in control*

This type of control means that any variation in output, which is not a result of variation in the input command, can be counterbalanced at any moment. The output can thus be linked only to the input command, whatever the external disturbance. The output is *servoed* by the input. Control is carried out in the *closed-loop mode*.

Any modification to the input command produces a reaction in the system, until an output consistent with the input is established. As shown in Figure 8.4, the output is not established until a certain number of oscillations are experienced (see Figure 8.5). The time taken for the output to stabilize (ie for the oscillations to disappear) is the *response time* of the system. It defines the instant at which the system no longer exhibits oscillatory behaviour or transient operation, and achieves stability or steady operation.

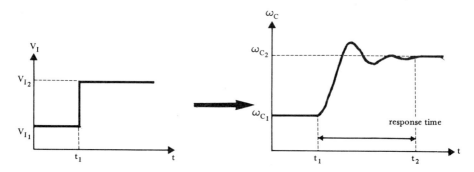

Figure 8.5. *Output oscillations*

8.1.2 CHARACTERISTIC PROPERTIES

8.1.2.1 Accuracy

In steady operation, there can be a difference in form between the final value of the output and the input. Since the system has been designed to control any difference between input and output, the difference must be caused by an element of the system not yet considered: the accuracy of the component parts. For example, the actuator must be supplied with a voltage greater than the minimum threshold value V_{min} ($V_{min} \neq 0$), if it is to be started. Then, in the system shown in Figure 8.3, any variation when $\Delta V_m < V_{min}$ will not allow compensatory movement of the actuator. Instead, it will cause an error $\Delta \omega_m$ which cannot be compensated for.

In order to minimize the error, the variation ΔV_m must be amplified. If an amplifier with gain K ($K > 1$) is placed before the differential amplifier, then ΔV is multiplied by K. To avoid restarting the actuator, it is important that:

$$K\Delta V \leqslant V_m \qquad (8\text{-}1)$$

which can involve an error in output speed such that:

$$K\Delta \omega = \Delta \omega_m \qquad (8\text{-}2)$$

so that:

$$\Delta \omega = \Delta \omega_m / K \qquad (8\text{-}3)$$

Equation (8-3) shows that under these conditions the error involved is divided by the value of the gain of the amplifier placed in front of the differential amplifier.

In a servo-system it is possible to reduce the error in steady operation (and thus to increase accuracy) by amplifying the value of the difference between the input and the output.

8.1.2.2 Stability

It has been pointed out already that transient operation in a servo-system corresponds to oscillatory behaviour. This phenomenon is due to closed-loop control, which allows a variation in input to be automatically developed in opposition to a variation in output.

The actuator and more generally the load it moves introduce inertia and friction producing a resistive torque that modifies the motor torque. These effects vary according to the energy supplied to the actuator (eg a function of speed).

Thus, if input to the servo-system is supplied with a voltage of large amplitude variation (eg a large ΔV for the servo-system shown in Figure 8.3), output will be subject to a sudden variation, which might prove excessive for the load.

This over-reaction is more marked when the level of energy supplied to the system is high. The oscillations that occur during transient operation will thus be greater in number when the stress on the system is greater, and the time required to dampen them will be longer. It may even happen, if the input is too strong, that the oscillations cannot be completely counteracted. In this case, the system can no longer stabilize itself, and is said to be *unstable*.

It has been shown that it is possible to amplify the difference between input and output values (ΔV). This amplification allows the amplitude of the energy supplied to the actuator to be reduced. If the gain of the amplifier is increased too much, the system can become unstable. There is a value for the gain, below which the system remains stable. This value is called the *critical gain*.

8.1.2.3 Stability versus accuracy

To summarize, by increasing the gain of the differential amplifier accuracy is increased and stability is reduced. Under these conditions it becomes difficult to choose between the need to increase the gain to attain a given level of accuracy and the need to reduce the gain so that the servo-system remains stable. These two imperatives constitute the 'stability-accuracy dilemma' often encountered in designing servo-systems.

8.1.2.4 Compensation

In attempting to find a solution to this problem, it should be noted that accuracy (or its inverse, error) affects only the stability of the system (ie the mode of operation that corresponds to small frequency variations), and that the stability only affects operation in which the variations are of high frequency. Therefore, it is necessary to attain at the same time a sufficiently large gain to satisfy accuracy requirements, and a sufficiently small gain to ensure stability. This can be achieved by applying only small amplification during transient operation, and greater amplification once steady operation commences. To carry this out successfully, an amplifier or attenuator in which the gain is varied depending on the frequency of the amplified signal is required (see Figure 8.6). This type of system is known as a compensation network. and most commonly comprises RC networks with a constant gain amplifier.

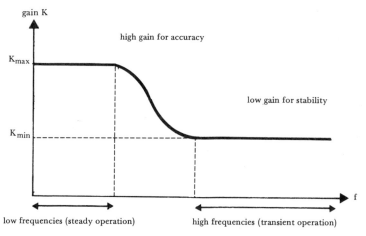

Figure 8.6. *Use of the amplifier in which the gain is varied relative to the frequency of the amplified signal*

Compensating a servo-system involves using a compensation network, but in certain cases, another approach can be employed. It has been shown that the stability of the servo-system depends on the level of energy supplied to the actuator (ie on the amplitude of the input and on the time required to transmit this variation). For a servo-system in which the input takes the form of motion (positional servo-system), energy is measured by the speed control of the actuator.

Instead of modifying the gain of the amplifier, as before, the gain corresponding to the accuracy required can now be maintained, but the braking effect due to feedback on the actuator increases with speed. Hence it is only necessary to subtract a quantity proportional to speed from the input. This is called *speed compensation*, or more commonly *tachometric compensation*, because the tachometric generator is used as a speed sensor (ie it provides a voltage proportional to angular speed).

8.1.3 MAIN TYPES OF SERVO-SYSTEM

Servocontrol of robot actuators acts on three main variables:

— movement (angular or linear);
— speed (angular or linear);
— torque (or force).

Thus, the following systems are used:

— a *positional* servo-system (see Figure 8.7);
— a *speed* servo-system (see Figure 8.8);
— a *torque* servo-system.

Figures 8.9 and 8.10 show the principles of operation of positional servo-systems compensated by RC networks and a tachometric generator respectively. These diagrams show only the basic structure. Calculating and correcting the difference can be carried out as easily using analog devices (differential amplifiers) as well as logical devices (eg microprocessors). If logical devices or microprocessors are used and the sensors are analog, devices capable of transforming analog values into digital values (or vice versa) must be used. These systems are called analog-digital converters (ADC) or digital-analog converters (DAC).

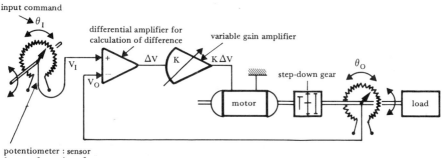

Figure 8.7. *Positional servo-system*

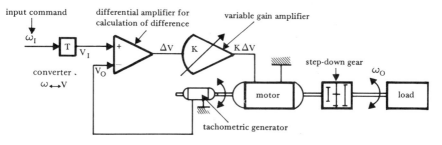

Figure 8.8. *Speed servo-system*

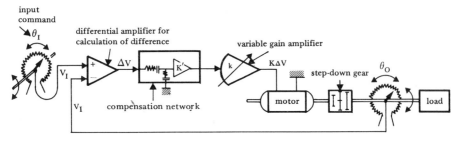

Figure 8.9. *Positional servo-system compensated by RC networks*

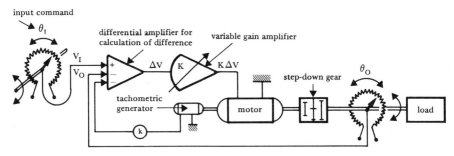

Figure 8.10. *Positional servo-system compensated by a tachometric generator*

Figure 8.11 shows the operating principle of a servo-system controlled by a microprocessor (which can in this case be made up of a single integrated circuit).

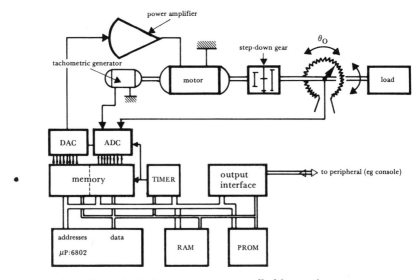

Figure 8.11. *Servo-system controlled by a microprocessor*

8.2 Mathematical stud of a servo-system

8.2.1 ESTABLISHING A SYSTEM EQUATION

In the general system shown in Figure 8.12, the following relationship exists:

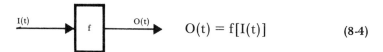

$$O(t) = f[I(t)] \qquad (8\text{-}4)$$

Figure 8.12. *Establishing a system equation*

However, if the system is servocontrolled, equation (8-4) is expressed as a differential, linear equation with constant coefficients:

$$b_0 O(t) + b_1 \frac{dO(t)}{dt} + b_n \frac{d^n O(t)}{dt^n} = a_0 I(t) + a_1 \frac{dI(t)}{dt} + a_m \frac{d^m I(t)}{dt^m} \qquad (8\text{-}5)$$

To find the solution $O(t)$, the differential equation (8-5) must be solved. The calculations can be simplified by introducing the Laplace transform, a mathematical transformation that associates each function $O(t)$ with a function $X(p)$ such that:

$$X(p) = \int_0^\infty x(t) I^{pt} dt \qquad (8\text{-}6)$$

In equation (8-5) this has the effect of modelling all terms of the form $d^n x(t)/dt^n$ into $p^n X(p)$, and thus transforming the differential equation into an algebraic equation:

$$b_0 O(p) + b_1 p O(p) + b_n p^n O(p) = a_0 I(p) + a_1 p I(p) + a_m p^m I(p) \qquad (8\text{-}7)$$

Thus:

$$O(p) = \frac{a_0 + a_1 p + a_2 p^2 + a_m p^m}{b_0 + b_1 p + b_2 p^2 + b_n p^n} I(p) \qquad (8\text{-}8)$$

It is now relatively easy to calculate $O(p)$. The inverse relationship is applied to return to the original $O(t)$:

$$O(t) = \frac{1}{2\pi j} \int_{C-j\infty}^{C+j\infty} O(p) I^{pt} dp \qquad (8\text{-}9)$$

8.2.2 THE TRANSFER FUNCTION

Equation (8-8) shows that if the initial conditions are equal to zero:

$$O(p) = F(p) I(p) \qquad (8\text{-}10)$$

The function $F(p)$ that characterizes the relationship between $O(p)$ and

I(p) is the *transfer function* of the system.

The transfer function allows the system to be 'represented' by a 'black box' or 'block' which ensures the relationship (equation 8-10) between O(p) and I(p). Consequently, the simple properties of association of the blocks can be deduced. Thus, the association of the two blocks shown in Figure 8.13 can be equivalent to a single block. The transfer function can be easily calculated, thus:

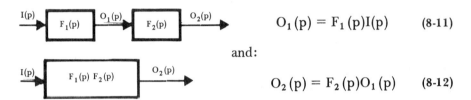

$$O_1(p) = F_1(p) I(p) \qquad (8\text{-}11)$$

and:

$$O_2(p) = F_2(p) O_1(p) \qquad (8\text{-}12)$$

Figure 8.13. *The transfer function*

so:

$$O_2(p) = [F_1(p) F_2(p)] I(p) \qquad (8\text{-}13)$$

The properties attached to these blocks allow the various sub-sections of the system to be represented in block schematic form, and can easily lead to a block by block calculation of the overall transfer function of the system. Figure 8.14 shows the block schematic equivalent of Figure 8.7.

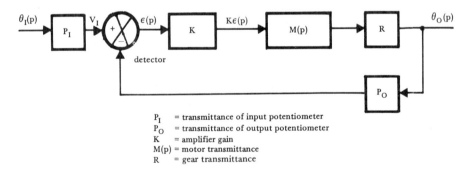

P_I = transmittance of input potentiometer
P_O = transmittance of output potentiometer
K = amplifier gain
$M(p)$ = motor transmittance
R = gear transmittance

Figure 8.14. *Block schematic equivalent of the system in Figure 8.7*

Equation (8-8) can be expressed in the form:

$$\frac{O(p)}{I(p)} = \frac{a_0}{b_0} \left[\frac{1 + \frac{a_1}{a_0} p + \frac{a_2}{a_0} p^2 + \frac{a_m}{a_0} p^m}{1 + \frac{b_1}{b_0} p + \frac{b_2}{b_0} p^2 + \frac{b_n}{b_0} p^n} \right] \qquad (8\text{-}14)$$

so:

$$\frac{O(p)}{I(p)} = KG(p) \tag{8-15}$$

Equation (8-15) expresses the most common transfer function, showing the static gain K, which characterizes accuracy and stability. If the system in Figure 8.12 is a loop system, such as in Figure 8.15, KG(p) is the transfer function in the open-loop mode.

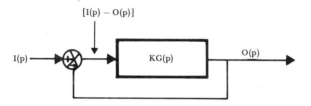

Figure 8.15. *Open-loop system*

The transfer function in the closed-loop system can easily be calculated by expressing the relationships between O(p) and I(p):

$$O(p) = KG(p)[I(p) - O(p)] \tag{8-16}$$

$$O(p)[1 + KG(p)] = KG(p)I(p) \tag{8-17}$$

Thus:

$$\frac{O(p)}{I(p)} = \frac{KG(p)}{1 + KG(p)} \tag{8-18}$$

8.2.3 QUANTITATIVE ANALYSIS OF SERVO-SYSTEM PERFORMANCE

8.2.3.1 Stability

If the initial conditions for the servo-system are taken into account, equation (8-7) should be written as:

$$\frac{(b_0 + b_1 p + b_n p^n)}{D(p)} [O(p) + I_1(p)]$$

$$= \frac{(a_0 + a_1 p + a p^m)}{N(p)} [I(p) + I_2(p)] \tag{8-19}$$

so:

$$O(p) = \frac{N(p)}{D(p)} \left[I(p) + \frac{I(p)}{D(p)} \right] \tag{8-20}$$

where $I_1(p)$ and $I_2(p)$ are defined by the initial conditions and $I(p) = I_2(p) - I_1(p)$. By applying the inverse transformation:

$$O(t) = \mathcal{L}^{-1}\left[\frac{N(p)}{D(p)}I(p)\right] + \mathcal{L}^{-1}\left[\frac{I(p)}{D(p)}\right] \qquad (8\text{-}21)$$

it can be shown that for the system to be stable, it is necessary and sufficient that $\mathcal{L}^{-1}\left[I(p)/D(p)\right]$ should tend towards zero when t tends towards infinity. This condition is fulfilled if the roots of the characteristic equation:

$$D(p) = 0 \qquad (8\text{-}22)$$

have negative real parts.

There are mathematical criteria expressed according to the coefficients of equation (8-22), which allow the conditions for which equation (8-22) only has roots with negative real parts to be determined. These conditions allow in particular the critical gain K_c to be calculated.

8.2.3.2 Frequency response

By *frequency response* is meant the permanent response of a system to a sinusoidal input. This frequency response can be found directly from the sinusoidal transfer function, that is the transfer function in which the variable p is replaced by the complex frequency $j\omega$. The sinusoidal transfer function is thus a complex function characterized for each value of ω by its modulus and argument. It can be represented in three different ways:

1. for each value of ω, each point of the curve is defined in the complex plane by the modulus and the argument of the transfer function: this is the *Nyquist locus*;
2. for each value of ω, the curve of the modulus (expressed in decibels) and the curve of the arguments (expressed in degrees) are traced separately; the scale of the abscissa (carrier) is logarithmic: this is the *Bond locus*;
3. for each value of ω, there is a corresponding point, defined by Cartesian coordinates carrying the abscissa of the argument (expressed in degrees) and in ordinates, the modulus (expressed in decibels): this is the *Black locus*.

The frequency response curves allow the following values to be determined for each value of ω:

1. the *modulus*: that is the ratio of the amplitude of the output to the amplitude of the input; this is approximately the amplification or attenuation of the input command when passing through the system;
2. the *argument*: that is the phase difference between the output signal and the input command or the delay of the input command when passing through the system.

It is possible to determine certain significant parameters from the curves, such as:

1. the *static gain* or the value of the modulus when $\omega = 0$ (the static gain characterizes accuracy and stability);
2. the *pass band* or the frequency band when the reduction in amplitude of the modulus is less than a given value; it measures the capacity of a system to respond to rapid commands, and thus indirectly defines its response time;
3. the *resonance factor* which measures the maximum value of the modulus for a given frequency (known as the *resonance frequency*); it allows the quality of transient operation and particularly the damping capacity of the system to be defined.

Each representation corresponds to a method for analysing the stability of the system and measuring its various characteristic parameters.

8.2.3.3 Compensation

Compensating a system involves inserting into the block schematic system a block (either in series or parallel) with a transfer function corresponding to the compensation network chosen. The modifications introduced must be defined using the methods mentioned above, particularly the frequency response curves.

8.3 Specific practical problems involved in the use of a robot servo-system

For each servo-system corresponding to a degree of freedom of the robot, the movements must take place at a given speed. It is important that the servo-system can be controlled both for speed and position. This causes practical problems since it involves combining the two systems shown in Figure 8.7 and 8.8. This cannot always be carried out. On the other hand, it is true that:

1. the servo-system moves at a given, controlled speed as long as the difference $(\theta_I - \theta_O)$ does not reach $\Delta\theta_M$;
2. the servo-system is regulated in position when the difference is less than $\Delta\theta_M$.

Under these conditions, the only practical problem is switching from the speed servo-system to the positional servo-system when the difference is equal to $\Delta\theta_M$.

8.3.1 ANALOG CONTROL SERVO-SYSTEMS

8.3.1.1 Switch system

Figure 8.16 shows a possible method for switching analog control of a servo-system. θ_I is the positional command, $V_{\omega I}$ is the voltage corresponding to the speed command and ΔV_M is the voltage corresponding to the threshold $\Delta\theta_M$ at which the positional servocontrol loop is switched.

8.3.1.2 System without switching

There is another system which does not involve a switch. A positional servo-system following the principle shown in Figure 8.10 is formed where the polarization of the amplifier defines the speed of the command ω_I. The resulting performance is shown in Figure 8.17. When the difference is greater than ΔV_M, speed is constant, since the amplifier is saturated. As soon as the difference is less than ΔV_M, the system operates by regulating the position. Under these conditions, speed is not regulated, since the system operates in the open-loop mode when the amplifier is saturated. This solution is convenient because of its simplicity, but is not suitable for high performance robots, particularly for large movements.

8.3.2 DIGITAL CONTROL SERVO-SYSTEMS

Switching is easily achieved when the system is under computer control. It involves no more than a simple algorithm. The operational basis of the servo-system is the same as that shown in Figure 8.11, but the program is changed. Figure 8.18 shows a typical algorithm for digital control.

8.3.2.1 Direct control without a converter

It is possible to control a continuous actuator directly with a digital signal. Under these conditions, the energy supplied to the actuator is proportional to the frequency of the input. This solution avoids the conversion stages but control of the servo-system is sensitive. Although this type of control offers many advantages, it requires considerable experience of electro-technical systems, electronics and data-processing to be used successfully. Some of the problems involved can be overcome by using a processor for each servo-system.

Actuator servocontrol

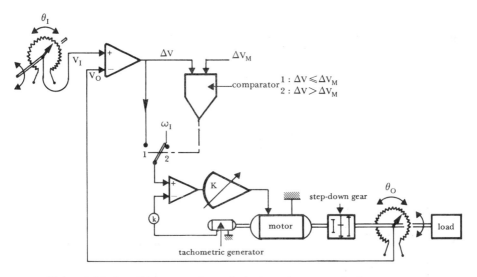

Figure 8.16. *Possible system for switched, analog control of a servo-system*

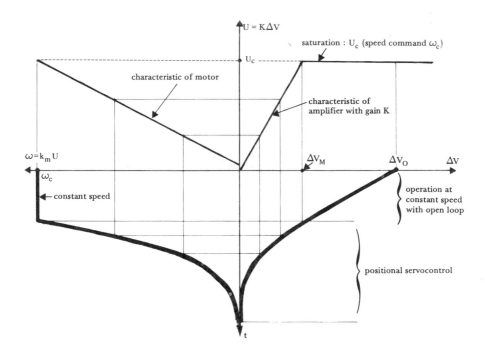

Figure 8.17. *Operation without switching*

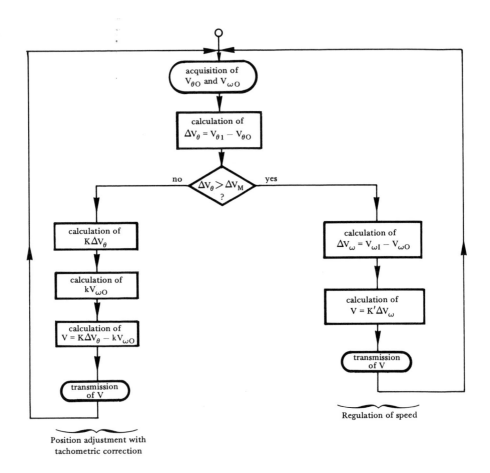

Figure 8.18. *Algorithm for digital control of a servo-system*

Chapter 9
Robot actuators

The motors that create motion in the robot articulations are called *actuators*. There are three main types of actuator, each associated with a different form of energy: *pneumatic*, *hydraulic* and *electrical*.

Pneumatic energy is most commonly used in the form of compressed air at pressures lower than 10 bars. It is easy to use, and is transmitted through flexible tubes, but since it does not provide in itself any lubrication, its use can result in significant friction in the moving parts. Moreover, compressed air almost invariably contains water vapour, which can result in oxidation. Compressed air is noisy, and servocontrolled pneumatic systems are difficult to regulate, which limits their mode of operation to being on or off, and thus to use with fixed or variable sequence robots (Classes 2 and 3 in the JIRA classification, see Chapter 2). These robots are the most commonly used and the least sophisticated.

Water is rarely used as the transmission fluid in hydraulically powered systems. Mineral oils of varying viscosity are most often used. The oil pressure does not usually exceed 100 bars in industrial robots both for safety reasons and to allow the use of flexible tubing. Although hydraulically powered systems provide a better power-to-weight ratio than pneumatically or electrically powered systems, hydraulic energy introduces a number of practical problems involving:

1. transmission of oil under pressure to the moving parts;
2. filtration of the oil to eliminate particles larger than 5 μm, which erode the narrow sections of the tubing used for controlling flow rate etc;
3. elimination of air;
4. the need for a central distribution unit, which increases the cost of hydraulic systems compared with systems using pneumatic or electrical energy;
5. maintenance, which requires specialized skills.

All these points restrict the use of hydraulic energy to robots with *high load capacity*. Some years ago the lower limit for a high load capacity was 30 kg. Today it falls between 50 and 100 kg.

Electrical energy is ideally suited to the control of robots, since it is easily available, is non-polluting and can be transmitted through cables.

The power-to-weight ratio in electrical systems is, however, lower than that in the other two systems. This is the reason behind the research being carried out into motors of high torque and low weight. Less than half of all industrial robots are electrically actuated, since sequence robots, which are the most common, are pneumatically powered and robots with high load capacity are hydraulically powered. The future of robotics, and the development of second and third generation robots will depend far more on electrical energy than on the other two forms.

9.1 Pneumatic actuators

9.1.1 BASIC PRINCIPLES

Actuators which make use of a fluid under pressure (ie pneumatic and hydraulic actuators) use potential energy stored in the fluid. Consider a length of tubing of cross-section S. The thrust (f) on a mobile body dividing the length of tubing into two parts in which the pressures p_1 and p_2 are respectively maintained, will be:

$$f = S(p_1 - p_2) \tag{9-1}$$

If the mobile body moves a distance dx, the work carried out can be expressed as:

$$fdx = S(p_1 - p_2)dx \tag{9-2}$$

and if the body moves distance dx in time dt, the transformation of the fluid power (right-hand term in equation 9-2) into mechanical power is described by:

$$\frac{fdx}{dt} = S(p_1 - p_2)\frac{dx}{dt} \tag{9-3}$$

Although equation (9-3) refers to a body moving in a straight line, it is also valid for a rotating body. In this case, another term C, representing the couple, is considered:

$$\frac{Cd\theta}{dt} = V(p_1 - p_2)\frac{d\theta}{dt} \tag{9-4}$$

where V is volume and θ is the angle of displacement.

The main difference between the use of air and oil is the high compressibility of air, which gives rise to complex phenomena.

9.1.2 PNEUMATIC SYSTEMS

The pneumatic system is made up of the following components:

1. the motor itself, normally a piston when used in robots;
2. the distributor, which controls the motor at appropriate intervals.

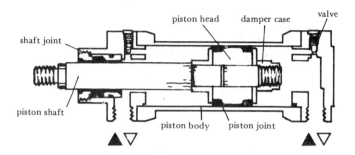

Figure 9.1. *Pneumatic linear piston*

Figure 9.1 shows a *linear piston*. This is a piston which moves from one limit of its stroke to the other depending on the direction of the air that drives it. Adjusting the air flow allows damping of the piston shaft at its limit. *Rotary pistons* are also available, but these are essentially linear pistons in which the piston shaft includes a rack which drives a pinion (known as a *rack and pinion system*).

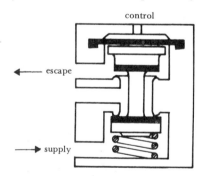

Figure 9.2. *Pneumatic valve distributor*

There are several types of distributor. The most commonly used are those with valves (see Figure 9.2) and those with spools (see Figure 9.3).

In the *valve distributor*, the valve is moved by a membrane which is subjected to the action of a control signal, and returned to its original position by a spring.

In the *spool distributor*, the spool moves across ports covering and uncovering them, thus diverting the air through different apertures.

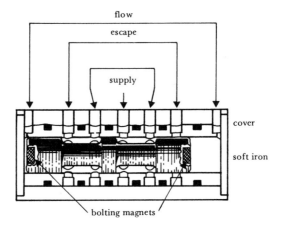

Figure 9.3. *Pneumatic four-track spool distributor*

9.2 Hydraulic actuators

As with pneumatic systems, hydraulically controlled systems are made up of the actuator itself (ie linear or rotary pistons in robots) and the distributor. The major difference lies in the fact that the distributor can be *proportionally controlled*, thus allowing the conventional use of hydraulic servo-systems, whereas in pneumatic systems the distributor operates in the on-off mode.

9.2.1 LINEAR PISTONS

There are three types of linear piston: the *single-action piston*, the *double-action piston* and the *differential piston*.

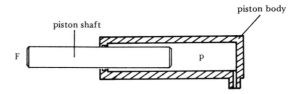

Figure 9.4. *Hydraulic single-action piston*

In the single-action piston (see Figure 9.4), the force developed is unidirectional. A return device (eg a spring) ensures the return of the piston shaft to its starting position. The double-action piston has two chambers in each of which pressures p_1 and p_2 can be established in turn (see Figure 9.5). It should be noted that the presence of the piston shaft in only one chamber means that the piston is asymmetrical with

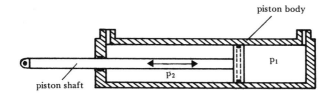

Figure 9.5. *Hydraulic double-action piston*

respect to the pressures needed to obtain the same movement to the right and to the left.

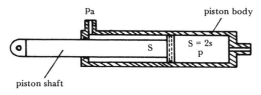

Figure 9.6. *Hydraulic differential piston*

For long strokes the differential piston is used (see Figure 9.6). Cross-section s of the piston shaft is equal to half the surface area of the piston body S, ie S = 2s. The force developed is:

$$f = s(2p - p_a) \qquad (9\text{-}5)$$

This piston requires only one three-track sluice (the double-action piston requires four tracks).

9.2.2 ROTARY PISTONS

As is the case with pneumatic pistons, it is possible to transform the linear motion of a hydraulic piston into rotary motion by mechanical means. There is, however, a type of piston designed to rotate. This is the *flapper piston* (see Figure 9.7). The simplest version is made up of a

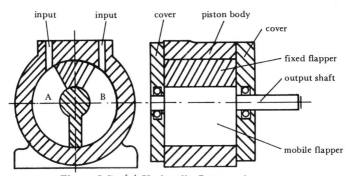

Figure 9.7. *(a) Hydraulic flapper piston*

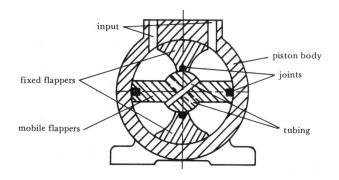

Figure 9.7. *(b) Hydraulic flapper piston in cross-section*

cylindrical body with a fixed flapper. Inside this pivots a mobile flapper which is integral with the output shaft. The stroke is limited, by the dimensions of the flappers, to about 330 degrees.

Since hydraulic pumps are not, strictly speaking, used as motors for robots the subject is not described further, and the reader is advised to consult specialized works on the subject.

9.2.3 SPOOL DISTRIBUTORS

A typical example of a *four-track distributor* is shown in Figure 9.8. The role of the distributor is to measure the flow q entering a chamber of known pressure $[(p_a + p_b)/2]$. As shown in Figure 9.8, the spools of length d adapt to a port of length d', and three different situations may arise: $d = d'$ (distributor without leakage or recovery), $d > d'$ (distributor with recovery) and $d < d'$ (distributor with leakage).

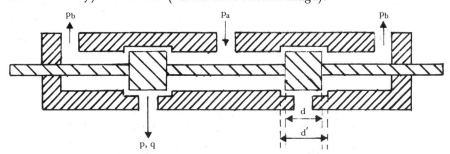

Figure 9.8. *Hydraulic four-track distributor*

9.2.4 SERVOVALVES

The *servovalve* is a device which moves the spool of a distributor in such a way as to uncover sections, adjusting the rates of flow proportional to the electrical control signal. This is essential for hydraulic servocontrol. Figure 9.9 shows a servovalve. It is a complex device characterized by many parameters. The most important parameters are:

1. the gain in flow or the relationship between the flow through the valve and the control current with no load (operating apertures short-circuited);
2. the gain in pressure or the relationship between the difference in output pressure and the control current when the apertures are closed;
3. the flow-pressure curves or the relationship between flow and the difference in pressure when the current is constant.

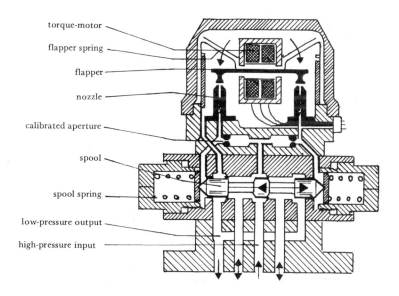

Figure 9.9. *Hydraulic servovalve*

9.3 Servocontrolled hydraulic systems

Consider the system made up of a servovalve controlled by current I, which moves its distributor by U, thus bringing about the movement x of a linear piston. Taking into account the geometrical complexities and imperfections, and the dynamic properties of the fluid, not one of the components has a simple, linear model. In practice, it is necessary to make approximations, for example, expressing the relationship between the current I used to control the servovalve and the movement U of the distributor as a second-order transfer function:

$$T_1(p) = \frac{U(p)}{I(p)} = \frac{K_1}{1 + 2\xi_1 \dfrac{p}{\omega_1} + \dfrac{p^2}{\omega_1^2}} \qquad (9\text{-}6)$$

where K_1 is a gain, ξ_1 is the damping effect close to unity and ω_1 is a natural pulsatance.

In the same way, the relationship between piston movement x and distributor movement can be represented as a transfer function:

$$T_2(p) = \frac{K(p)}{U(p)} = \frac{K_2}{p(1 + 2\xi_2 \dfrac{p}{\omega_2} + \dfrac{p^2}{\omega_2^2})} \qquad (9\text{-}7)$$

where ω_2 is smaller than ω_1.

The overall transfer function will be:

$$T(p) = \frac{x(p)}{I(p)} = T_1(p) T_2(p) \qquad (9\text{-}8)$$

This is a standard system, which functions as an open loop. Figure 9.10 shows this system.

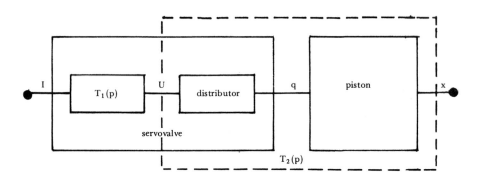

Figure 9.10. *Hydraulic servovalve and piston*

Servocontrol can be carried out mechanically by associating the piston with a distributor mounted in such a way as to make the admission openings into the chambers of the piston depend on the position of the spool relative to that of the load (see Figure 9.11). Adding an electrical motor drive creates an *electro–hydraulic system* with positional servocontrol (see Figure 9.12).

Hydraulic systems (eg pistons, motors, distributors, servovalves) can vary widely. There are many types of hydraulic control unit. Two examples are given in Figures 9.11 and 9.12.

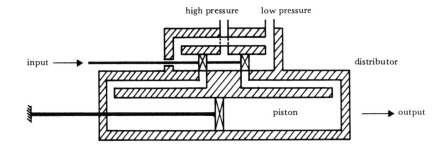

Figure 9.11. *Mechanical servocontrol system*

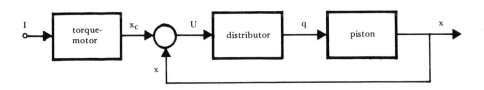

Figure 9.12. *Electro-hydraulic system with servocontrol*

9.4 Electrical actuators

Electrical control units do not have the obvious advantages of pneumatic and hydraulic systems. Servovalves and distributors, for instance, require mechanical precision to ensure adequate operation, which are difficult to servocontrol, and are connected by thick, inflexible tubes with risks of leakage at the joints. Against these advantages must be weighed the major drawback of electrical systems: the low power-to-weight ratio, which is a serious disadvantage for robots with integral motors, or transmission systems, which cause other problems. Any type of motor can be used, but at present only two types are used in robots: the *direct current (d.c.) motor* and the *stepping motor*.

Asynchronous motors are not suited for robots at varying operating speeds in both directions of rotation. Only the somewhat unresearched self-controlled synchronous motors are suitable as robot actuators. Linear motors do not provide a satisfactory level of performance.

9.4.1 DIRECT CURRENT MOTORS

These have the advantage of providing torque independent of the position and speed of the motor. They include two parts:

1. a fixed inductor made up of a coil through which an induction current I_1 flows, or of magnets;

2. an armature which is a mobile coil through which a current I_2 flows.

The magnetic fields created by the inductor and the armature are in quadrature, and thus have no influence over each other.

The electromotive force E is proportional to the speed of rotation ω and the flux ϕ created by the inductor:

$$E = K\omega\phi \qquad (9\text{-}9)$$

During steady operation (see Figure 9.13):

$$V_1 = R_1 I_1 \qquad (9\text{-}10)$$

$$V_2 = E + R_2 I_2 \qquad (9\text{-}11)$$

The electromagnetic torque C_m will be such that:

$$C_m \omega = E I_2 \qquad (9\text{-}12)$$

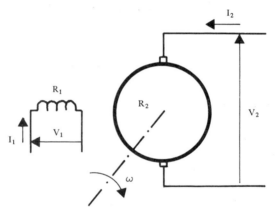

Figure 9.13. *Direct current motor in steady operation*

The d.c. motors used in robotics generally have permanent magnets in the inductor. There are three main types:

1. *standard motors* in which the armature is wound onto a magnetic material (see Figure 9.14);
2. *bell motors* in which the armature conductors are attached to an insulated cylinder (see Figure 9.15);
3. *disk motors* in which the armature conductors are attached to or wound onto an insulated disk (see Figure 9.16a and b).

In the standard and bell motors, induction is radial and the armature currents are axial. Induction is axial and the currents are radial in the disk motor.

When permanent magnets are used as inductors, it is ideal to combine

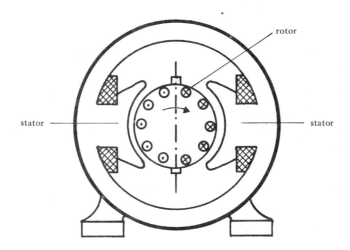

Figure 9.14. *Cross-section of the standard direct current motor*

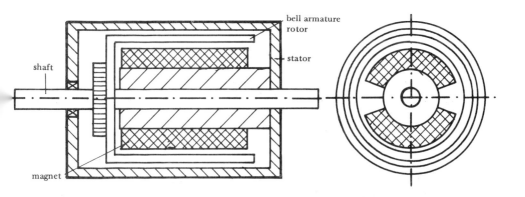

Figure 9.15. *Motor with bell rotor showing arrangement of magnets*

lightness, a high level of induction and stability of induction with variations in temperature. As yet there is no ideal magnetic material, but the three types of permanent magnet currently in use are:

1. *Alnico magnets* (ie an alloy containing varying amounts of iron, aluminium, cobalt, nickel and copper; induction is of the order of 1 Tesla);
2. *ferrite magnets* (ie compounds of powdered iron and barium or strontium oxides; induction is of the order of 0.5 Tesla);
3. *rare earth and cobalt magnets* (eg Samarium-cobalt). These give the best results, but are costly.

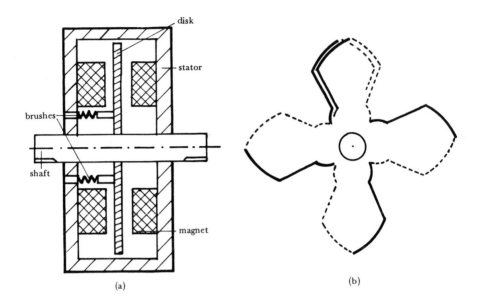

Figure 9.16. *Disk motor: (a) cross-section, (b) coil shape*

It is important to note that these motors are rarely used in steady operation, but more commonly in the transient mode when used to activate robot articulations. It is from this point of view that a motor should be judged with particular regard to various time constants.

In Figure 9.17 the speed of rotation ω is plotted against the torque C, for which theoretically there should be no limit, but there are limits resulting from the existence of certain physical phenomena. Increase in temperature brought about by various Joule effect losses in the coils is a limitation. The maximum temperature tolerated by the system is 150°C, which is quickly reached.

The operational limit can be extended (see Figure 9.17) by ventilating the motor adequately or even cooling it with circulating water or oil. If the problem of overheating is solved, the next limitation arises from commutation, which dictates that the current supplying the coils must periodically change direction. If the speed increases sparking will occur in the commutator. This is also a limiting factor.

Finally, even if the problems of overheating and commutator sparking are overcome, the mechanical structure of the motor itself cannot withstand a considerable increase in ω, because of the centrifugal force and vibrations caused by imbalance etc. Three such limitations are shown in Figure 9.17.

Robot actuators

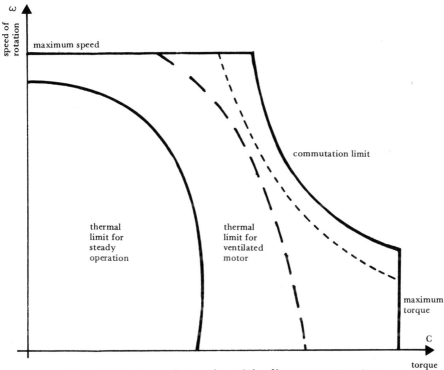

Figure 9.17. *Areas of operation of the direct current motor*

9.4.2 STEPPING MOTORS

The stepping motor is synchronous in that the input command and the position of the motor are synchronized. Although the structure can be varied, especially if a large number of steps per revolution is required, the basic principle is simple.

Consider the motor shown in Figure 9.18. The stator is made up of four poles with coils. The rotor is a permanent magnet. If the coils $\alpha\alpha'$

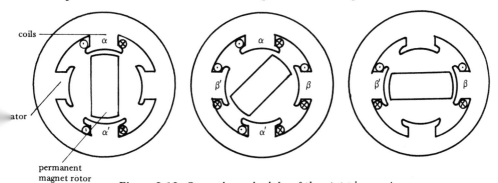

Figure 9.18. *Operating principle of the stepping motor*

are supplied with current, the rotor takes up a position in which the flux through $\alpha\alpha'$ will be maximum. If only $\beta\beta'$ are supplied, the rotor turns through 90 degrees. If α and β are supplied, the rotor takes up the mean position, and so on. If $\alpha\alpha'$ and $\beta\beta'$ are assumed to be identical coils, with identical current flow, then supplying $\alpha\alpha'$, then $\beta\beta'$, then $\alpha\alpha'$ while changing the direction of the current and then $\beta\beta'$ and so on allows the motor four stable positions at 90 degree intervals.

If the number of poles is increased on the stator as well as the rotor, the number of stable positions of the rotor can be greatly increased. This is analogous to variable reluctance. The number of stable positions can also be increased by placing several stacks in stable positions at appropriate intervals. In this way steps of a few minutes can be formed.

Stepping motors are controlled by pulses of current which cause the rotor to start, accelerate and stop when moving from one position to the next. There are certain problems, since the input command must take into account the inertia of the rotor at the expense of synchronicity. For robot control, a stepping motor with large torque is a solution, and one which avoids the use of a positional sensor. In practice, counting the steps can be a delicate operation as is synchronization of several motors actuating the DOF at different speed, and with variable levels of inertia and torque.

9.5 Servocontrolled electrical motors

In this section, those d.c. motors currently used in electrically powered robots will be discussed.

9.5.1 TRANSFER FUNCTIONS AND BLOCK SCHEMATIC EQUIVALENTS

The parameters of the servo-motor (see Figure 9.19) are as follows:

r: resistance;
l: inductance;
i_f: inductor current;
V_f: inductor voltage;
R_m: armature resistance;
L_m: armature inductance;
i_m: armature current;
V_m: armature voltage;
θ_m: position of armature;
ω_m: speed of armature rotation;
J_m: inertia of rotor providing torque Γ_m;
f_m: viscous friction of rotor;
k_m: torque coefficient;

θ_c: angular position of load;
ω_c: speed of rotation of load;
θ_m/θ_c: reduction ratio;
J_c: inertia of load;
f_c: coefficient of viscous friction of load;
k_c: return coefficient of load.

These parameters allow the transfer functions of the servo-motor to be calculated.

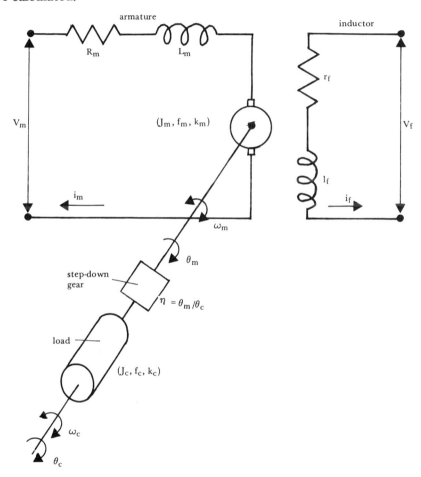

Figure 9.19. *Direct current motor with step-down gear moving load*

9.5.1.1 Induction control motors

The following equations can be established:

$$V_f = r_f i_f + l_f \frac{di_f}{dt} \tag{9-13}$$

$$\Gamma_m = k_m i_f \qquad (9\text{-}14)$$

$$\Gamma_m = J\frac{d^2\theta_m}{dt^2} + F\frac{d\theta_m}{dt} + K\theta_m \quad \text{(mechanical equation)} \quad (9\text{-}15)$$

with:

$$J = J_m + \frac{J_c}{\eta^2} \qquad (9\text{-}16)$$

$$F = f_m + \frac{f_c}{\eta^2} \qquad (9\text{-}17)$$

$$K = \frac{k_c}{\eta^2} \qquad (9\text{-}18)$$

being respectively, total inertia, total viscous friction and total feedback relating to drive shaft. If the Laplace transform is introduced, the three equations become:

$$V_f(p) = (r_f + l_f p)I_f(p) \qquad (9\text{-}19)$$
$$\Gamma_m(p) = k_m I_f(p) \qquad (9\text{-}20)$$
$$\Gamma_m(p) = (Jp^2 + Fp + K)\theta_m(p) \qquad (9\text{-}21)$$

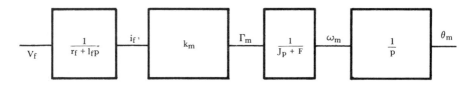

Figure 9.20. *Open-loop direct current motor (with load) with induction control*

The block schematic equivalent is shown in Figure 9.20. The transfer function is:

$$\frac{\theta_m(p)}{V_f(p)} = \frac{k_m}{(r_f + l_f p)(Jp^2 + Fp + K)} \qquad (9\text{-}22)$$

In practice, it is often assumed that $K = 0$, so:

$$\frac{\theta_m(p)}{V_f(p)} = \frac{k_m}{p(r_f + l_f p)(Jp + F)}$$

$$= \frac{k_m}{r_f F} \cdot \frac{1}{p(1 + \frac{l_f}{r_f}p)(1 + \frac{J}{F}p)} \qquad (9\text{-}23)$$

$$= \frac{k_0}{p(1 + \tau_e p)(1 + \tau_m p)}$$

τ_e is the electrical time constant. It is negligible compared with the mechanical time constant τ_m. For this reason the transfer function of the induction control motor is often taken to be:

$$\frac{\theta_m(p)}{V_f(p)} = \frac{k_0}{p(1 + \tau_m p)} \quad (9\text{-}24)$$

or, if ω_m is the speed of rotation ($\omega_m = d\theta_m/dt$):

$$\frac{\omega_m(p)}{V_f(p)} = \frac{k_0}{(1 + \tau_m p)} \quad (9\text{-}25)$$

9.5.1.2 Armature control motors

The equations become:

$$V_m = R_m i_m + L_m \frac{di_m}{dt} + k_e \omega_m \quad (9\text{-}26)$$

$$\Gamma_m = k'_m i_m \quad (9\text{-}27)$$

$$\Gamma_m = J \frac{d^2 \theta_m}{dt^2} + F \frac{d\theta_m}{dt} + K\theta_m \quad (9\text{-}28)$$

Rotation gives rise to a back-electromotive force (k_e) proportional to angular speed ω_m.

Using the same method, the following relationship is found:

$$\frac{\theta_m(p)}{V_m(p)} = \frac{k'_m}{JL_m p^3 + (JR_m + FL_m)p^2 + (L_m K + R_m F + k'_m k_e)p + KR_m} \quad (9\text{-}29)$$

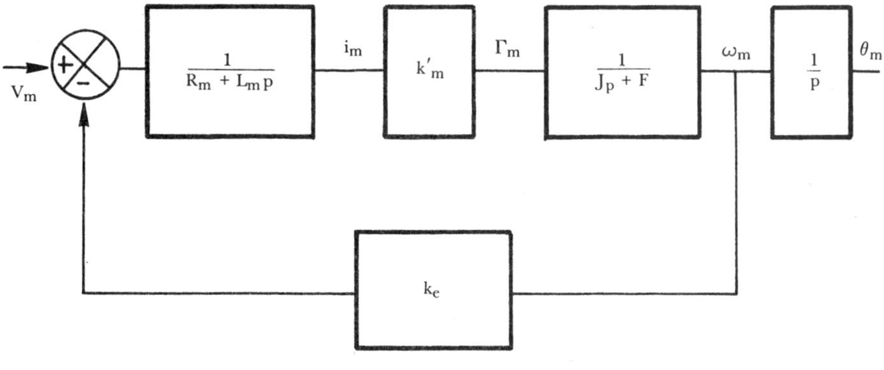

Figure 9.21. *Armature control direct current motor with velocity feedback*

If, as is generally the case, K = 0:

$$\frac{\theta_m(p)}{V_m(p)} = \frac{k'_m}{p[(R_m + L_m p)(F + Jp) + k_e k'_m]} \quad (9\text{-}30)$$

The block schematic diagram is shown in Figure 9.21.

9.5.2 TUNING THE SERVO-MOTOR

Consider a motor controlled by the inductor, situated in a servocontrol loop (see Figure 9.22), in which the desired value of the output θ_O is equal to input θ_I. To achieve this, θ_I will be fixed using a potentiometer circuit. θ_O will be measured using another potentiometer, and $\theta_I - \theta_O$ will feed the inductor after amplification.

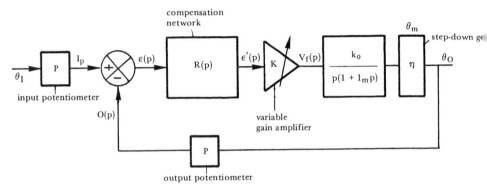

Figure 9.22. *Positional servocontrol loop with compensation network for the induction control motor*

In Chapter 8 it was shown that to ensure satisfactory servo-motor performance from the point of view of stability and accuracy, a compensation network must be introduced into the servocontrol circuit. More precisely, this consists of error compensation $\epsilon(t) = I(t) - O(t)$, by a proportional amount:

with $\epsilon(t)$ (proportional compensation)

with the derivative $\dfrac{d\epsilon(t)}{dt}$ (derivative compensation)

with the integral $\displaystyle\int_0^t (t)dt$ (integral compensation)

In practice, compensation combines at least two of these artefacts: *proportional and derivative compensation (PD)*, *proportional and integral compensation (PI)* and sometimes a combination of all three *(PID compensation)*.

9.5.2.1 Proportional and derivative compensation

This type of compensation network creates an error signal $\epsilon'(t)$, proportional to $\epsilon(t)$ and to $d\epsilon(t)/dt$, thus:

$$\epsilon'(p) = (k + \lambda p)\epsilon(p) \tag{9-31}$$

It can be formed electronically as shown in the circuit in Figure 9.23.

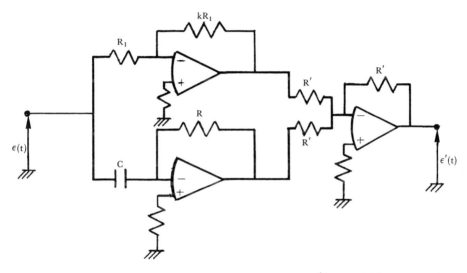

Figure 9.23. *Proportional and derivative compensation:* $\epsilon'(p) = \epsilon(p)[k + (RC)p]$

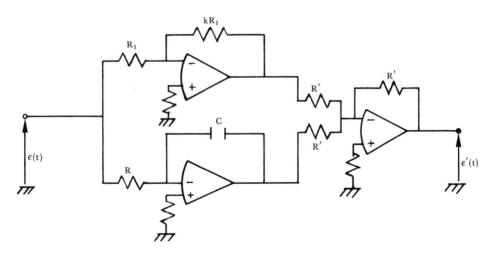

Figure 9.24. *Proportional and integral compensation network:*
$\epsilon'(p) = \epsilon(p)[k + 1/(RC)p]$

9.5.2.2 Proportional and integral compensation

This type of compensation is intended to produce an error signal $\epsilon'(t)$ proportional to $\epsilon(t)$ and to $\int_0^t \epsilon(t)dt$, so:

$$\epsilon'(p) = \left(k + \frac{\lambda}{p}\right)\epsilon(p) \qquad (9\text{-}32)$$

A PI compensation network is shown in Figure 9.24.

9.5.2.3 Three-term compensation

This type of network combines the three effects mentioned above, so:

$$\epsilon'(p) = \left(k + \lambda p + \frac{\lambda'}{p}\right)\epsilon(p) \qquad (9\text{-}33)$$

A PID compensation network is easily assembled as shown in Figure 9.25. In all three cases, the input resistance R' can be varied, allowing each of the effects to be measured according to the required level of performance.

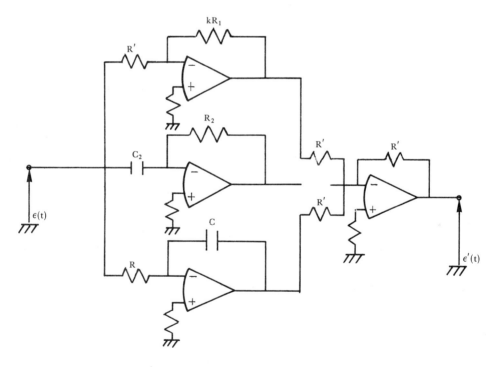

Figure 9.25. *Three-term compensation network:*
$\epsilon'(p) = \epsilon(p)[k + (R_2 C_2)p + 1/(RC)p]$

9.5.2.4 Tachometric compensation

In Chapter 8 it was shown that a system can be compensated using the voltage supplied by a tachometric generator. This is shown in Figure 9.26. It should be noted that the tachometric generator supplies a purely derivative compensative effect, which only takes the output into account in the form of parallel action.

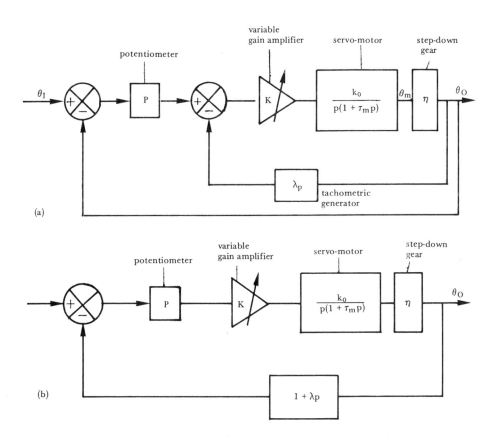

Figure 9.26. *Principle of the motor servocontrolled using a tachometric generator. The two feedback loops in (a) are equivalent to the feedback loop $(1 + \lambda p)$ in (b)*

9.6 Transmission systems

Between the shaft of the motor, where the active torque is developed, and the articulation which is to be moved are the transmission systems.

These systems have been little studied so far. Transmission systems can perform three things:

1. drive;
2. transform a rotational movement into a linear motion or vice versa;
3. reduce movements and amplify torque.

Transmission systems can be divided into two major categories according to their reversibility. This has a direct effect on robot control. In the irreversible transmission system (eg a ratchet system), supply to the actuators can be cut-off, and the robot will remain in the position it had attained at the moment of cut-off. In the reversible transmission system, robot position is affected by gravity as soon as the supply is cut. For this reason counterbalancing systems must be provided. These can either be mechanical systems (eg using parallelogram arrays) or electro-mechanical systems (eg brakes). On the other hand, reversibility has the advantage of transmitting the forces to which the robot is subjected back to the motor servo-systems, thus allowing robot-environment interactions.

The choice of transmission systems is determined by the nature of the actuators and the structure of the robot. The choice often comes down to the question: 'If the articulation to be moved is already set in a given position, and the motor is chosen, where can it be placed?' This problem is solved according to several criteria. The final criterion is how to connect the shaft of the motor to that of the articulation with satisfactory reduction ratio. There are many systems available, for example, chains, notched belts, round or twisted cables, metal bands, worm screws, connecting rods. The material used is of great importance, since the aim is to maximize strength whilst minimizing weight, inertia, flexibility and play, which are sources of vibration, and dry or viscous friction, which give rise to power losses.

Robot rigidity depends on the quality of the transmission systems, and rigidity is very important in industrial robots, since it affects precision. In some cases, the transmission system fulfils the role of a step-down gear, while in others, a step-down gear can be mounted on the motor shaft. This is the case with the harmonic drive where movement is transmitted in a ratio of one to one.

9.7 Conclusions

Actuators are essential to the robot, but a universal motor is not available for the articulated mechanical system. The motor would require modification to give the correct power-to-weight ratio and structure. Integrating step-down gears with the correct reduction ratio, minimum play, friction and inertia into the motor might be helpful.

If robot flexibility is the main characteristic of electrically driven robots and hydraulic motors can supply power at high levels, it is likely that the pneumatic actuator will continue to be used in fixed and variable sequence robots (see JIRA classification in Chapter 2).

Chapter 10
Internal sensors

A robot can be equipped with two types of sensor, *internal sensors*, which establish its configuration in its own set of coordinate axes, and *external sensors*, which allow the machine to position itself relative to its environment. This chapter is concerned with the most commonly used types of internal sensor, which measure movement, speed, acceleration and stress.

10.1 Movement or position sensors

There are many types of movement sensor, but only those commonly used are described here. Figure 10.1 summarizes the different types available.

Movements to be measured are either rectilinear (translational) or angular (rotational).

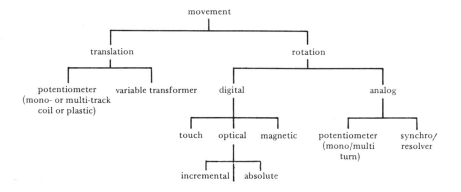

Figure 10.1. *Types of movement sensor*

10.1.1 RECTILINEAR MOVEMENT SENSORS

10.1.1.1 Potentiometer sensors

The most commonly used sensor is the *rectilinear potentiometer*, which can function on two different principles: the *coiled track potentio-*

meter and the *plastic track* or *hybrid potentiometer* (which is a standard coiled track covered with a plastic conductor).

The variation in resistance between the track end and the cursor allows the extent of the movement to be measured (see Figure 10.2). It is often necessary to alter the impedance between the cursor and its load, and this is achieved using an operational 'following' amplifier. The plastic track does not present the same problem as the coiled track when there is a discontinuity in passing from one loop to another; an infinitely variable impedance is obtained.

These potentiometers are also made with up to four tracks, and allow linearity of 0.2 per cent of the full scale to be achieved.

$$e = U \frac{R_2}{R_1 + R_2} \quad \text{for infinite load}$$

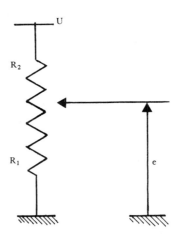

Figure 10.2. *Variation in resistance between the track end and the cursor*

Manufacturers allow for standard fixing with auto-aligning swivel joints or wheel disks. These potentiometers are made with a scale of movement ranging from 25 mm to 1,000 mm. At the top end of the range the systems are heavy and bulky, weighing as much as 2.5 kg.

Translational movement can be transformed into rotational movement, using a rack and pinion system and a rotary potentiometer.

10.1.1.2 Variable transformers

The sensor is made up of two fixed coils inside which is a mobile magnetic core on a shaft, mechanically linked to the movement to be measured. The movement of the core causes a variation in coupling between the coils. If the primary winding is fed with an a.c. supply, a voltage of the same frequency is detected in the second winding, but the amplitude is modulated by the position of the core. If these sensors are used, de-modulation, achieved electronically, is necessary, and a device for this purpose is sometimes incorporated into the sensor. In this system the range of movement varies from 0.6 mm to 38 mm.

10.1.2 ANGULAR MOVEMENT SENSORS

10.1.2.1 Potentiometer sensors

The *rotary potentiometer* is most commonly used. These potentiometer sensors work on the same principles as the rectilinear potentiometer and can achieve a high degree of linearity.

Although usable only on a limited number of turns, this is rarely an obstacle in practice. By placing two cursors diametrically opposite each other the potentiometer can be made mechanically continuous, although the output voltage between the two sections is non-linear, but of known value (see Figure 10.3).

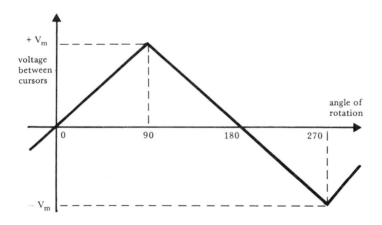

Figure 10.3. *Non-linear output of the potentiometer sensor*

Potentiometers can be assembled in several layers, with the control axes concentric. In this way, complex functions can be carried out, but problems of mechanical locking in the tracks become significant.

There is a wide range of models on the market, produced by a number of manufacturers.

10.1.2.2 Variable transformers

Similar to the linear movement sensor is the technique that uses variable transformers. Figure 10.4 shows two coils. The larger coil is fixed and the other coil is capable of rotation about an axis perpendicular to the plane of the figure.

If the inner coil is fed with an a.c. supply with voltage $U_1 = U\sin \omega t$, there will be voltage $U_2 = kU\cos \theta \sin \omega t$ induced between A and B, where θ is the angle between the axes of the two coils. This characteristic is common to two sorts of widely used angular sensors: the *synchro* and the *resolver*.

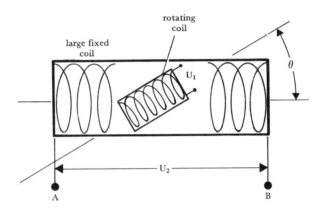

Figure 10.4. *Variable transformer*

Synchro: The synchro is made up of a stator of three coils, placed at an angle of 120 degrees to each other, at the terminals of which are voltages $kU\cos \theta \sin \omega t$, $kU\cos \theta \sin (\omega t + 2\pi/3)$ and $kU\cos \theta \sin (\omega t + 4\pi/3)$. These three modulated voltages allow θ to be determined (see Chapter 12). In servo-systems two identical systems mounted on a synchro detector are used often (see Figure 10.5):

1. the transmitter fixes the servo-system order by locking the rotor at value U_1;
2. the receiver locates the locking U_2 of the outgoing shaft of the body to be turned under servocontrol.

If the voltage applied to the transmitter rotor is $U_1 = U\sin \omega t$, a voltage $U_2 = kU\cos (\theta_1 - \theta_2)\sin \omega t$ is induced in the receptor rotor. This constitutes the error voltage.

In practice, the locked input and output values are located relative to the two unlocked axes of $\pi/2$ such that θ and ϕ are adjacent. Then $U_2 = kU\sin (\theta_1 - \phi)\sin \omega t = kU(\theta - \phi)\sin \omega t$.

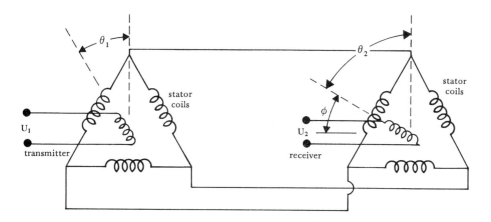

Figure 10.5. *Circuit diagram of the synchro*

Resolver: The resolver works on a principle similar to that of the synchro. The stator is made up of two fixed coils set at 90 degrees to each other. Synchros and resolvers can be used in digital coding systems such as are described in Chapter 12.

The a.c. output signals easily can be transmitted over long distances, even in noise, to remote control units. ASEA, for example, uses resolver movements as angular sensors in their IRB6 robot. Synchros and resolvers are extremely robust systems. The degree of accuracy attained by various models ranges from 7 to 20 minutes, and the frequency of utilization is of the order of 1 to 2 kilohertz.

10.1.2.3 Optical encoders

Optical encoders are angular sensors providing information on the angular position of the shaft on which they are mounted, using a binary code TTL. There are two types of optical encoder, *incremental* and *absolute* encoders.

Incremental encoders: The basic mode of operation is the same for all models, although it varies from one manufacturer to another. A photo-electric cell or a photoconductor detects variation in the pattern caused by the movement of a disk, on which are regularly spaced black lines, in front of a light source (see Figure 10.6). This alternating optical signal is transformed into a series of electrical impulses. Usually, incremental encoders have two main outputs, each generating a certain number of impulses per turn, up to 2×10^6. This number determines the accuracy of the sensor. The two output signals are de-phased by a quarter of a step, for example by unlocking the two tracks, so by examining the difference in phase between the two signals, the direction of rotation of

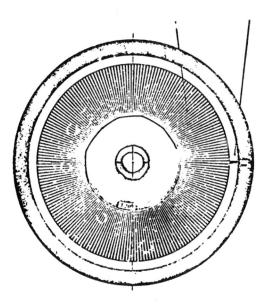

Figure 10.6. *Incremental disk*

the drive shaft can be determined. Moreover, a third output, known as the *marker*, produces a single impulse per turn and acts as a synchronizing signal. Figure 10.7 shows the typical output waveforms from these encoders. The direction of rotation is established using software, often provided by the manufacturer.

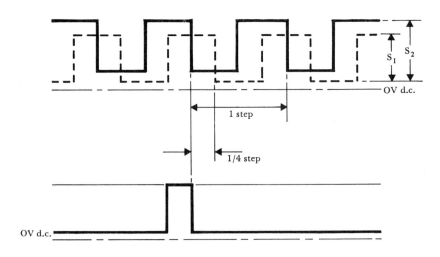

Figure 10.7. *Appearance of typical incremental encoder outputs*

Figure 10.8 is an example of the circuits used in the incremental encoder type MCB KG100. S_1 and S_2 are the two outputs of the encoder. The leading and trailing edges are detected and the rotational direction determined.

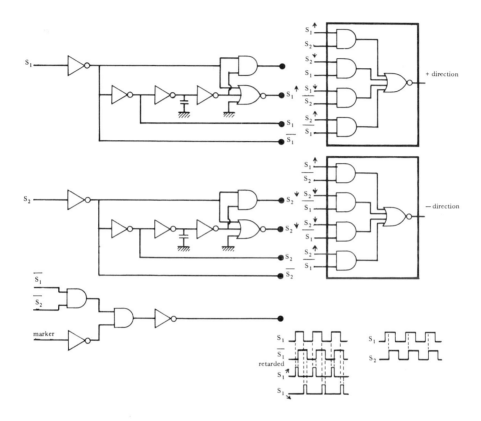

Figure 10.8. *Circuitry used for determining the rotational direction of an incremental encoder (MCB KG100)*

Usually incremental encoders are used in positional servocontrol, where the zero is not fixed. They are often used in impulse-generator systems for high-speed servocontrol. The frequency of one of these trains of impulses equals the product of the number of impulses per revolution and the speed of rotation (revolutions per second). If the frequency can be measured, the speed of the drive shaft can be calculated. In Chapter 12, the use of a microprocessor for these calculations will be discussed.

Absolute encoders: The absolute encoder takes the form of a disk on which black-and-white circles are arranged, so that on any radius the succession of black-and-white areas makes up a binary representation of the angle between the drive shaft and a known origin (see Figure 10.9).

Figure 10.9. *Cutting an absolute encoder disk*

The binary code is derived directly from the encoder disk, using an optical reading system similar to that used in the incremental encoder. The number of tracks on the disk determines the degree of resolution of the encoder.

The design of the tracks is chosen in such a way as to produce a Gray code or reflected binary code, in which the transition from one number to the following number is represented by the change of only 1 bit.

Figure 10.10 gives the binary and reflected binary codes for 3 bits. It can be seen that the transition from 3 to 4 is represented by a change of 3 bits in binary code, whereas the same transition in reflected binary code involves a single change. This technique is useful in avoiding errors and ambiguity in reading.

decimal	binary code $2^2\ 2^1\ 2^0$	reflected binary code
0	0 0 0	0 0 0
1	0 0 1	0 0 1
2	0 1 0	0 1 1
3	0 1 1	0 1 0
4	1 0 0	1 1 0
5	1 0 1	1 1 1
6	1 1 0	1 0 1
7	1 1 1	1 0 0

Figure 10.10. *Binary code and reflected binary code for 3 bits*

The manufacturer includes all electronic conversion equipment for this code in the encoder, in a pure binary code directly usable by a processor.

Using an absolute encoder allows the instantaneous value of the angle of rotation of the drive shaft to be found, relative to an initial locking of the encoder on the shaft. This is useful particularly when the equipment is under pressure. It is a great advantage when, by simply reading the encoders for each articulation, the servocontrol references can be adjusted to avoid sharp, possibly damaging, movements from the moment a robot starts to move.

A robot equipped with these sensors does not require a specific initialization position.

These sensors can be used easily and effectively with microprocessors.

Comparisons: Incremental or absolute encoders offer numerous advantages; the lack of contact allows them to turn very rapidly, as much as 5,000 revolutions per minute for a resolution of 100 points per revolution (resolution of 3.6 degrees).

The maximal speed varies inversely with definition, for example five revolutions per minute, for a resolution of 0.6. The lack of contact keeps drive torque very low.

These systems are difficult to use in noisy conditions, which is the case when cut-off supplies, or stepping motors are used. Moreover they are fragile, and may give inaccurate results when exposed to vibrations.

Incremental encoders are much less expensive than absolute encoders. These sensors are more widely used in robot technology.

10.2 Speed sensors

Speed sensors measure the speed of translational and rotational movements. In most cases, the calculation is confined to the speed of rotation, since to measure the speed of translation requires highly specialized sensors, which are rarely used.

When either a translation or rotation is measured by a potentiometer, the signal can be derived electronically, but this is not true for the speed sensor. This derivative can be calculated using a computer by taking positional samples with very small time intervals between them. Alternatively, counting can be used, as with incremental encoders. The number of impulses in a given time are counted. This method is not always satisfactory, especially near the limits of speed variation: at low speeds there is a risk of instability and at high speeds only a low level of accuracy is achieved.

This method, however, which has the advantage of using the same sensor for measuring position as for measuring speed, provides good control of speed, about a given operating point. This is true for all other pulse-generating speed sensors, such as toothed wheels.

Tachometric generators are without doubt the most commonly used speed sensors. They divide into two basic types, *direct current (d.c.) tachometric generators* or *tachometric dynamos*, and *alternating current (a.c.) tachometric generators*.

The first type is more widely used, and functions in the following way. It delivers a direct signal, directly proportional to the speed under surveillance. The signal is expressed in terms of volts/1,000 revolutions per minute. The choice of such a sensor is determined by its linearity (0.1 per cent can be attained), its level of residual oscillation, its maximum usable speed (3,000—8,000 revolutions per minute) and the inertia parameters. It is always helpful to mount a tachometric generator directly onto the connecting shaft, so that it can revolve at the highest possible speed within its range.

There are many models on the market.

Tachometric generators using an a.c. supply are far less frequently used, although suitable for remote control systems. Moreover, if used in conjunction with variable transformer positional sensors, the signals from each can be combined, so long as they are modulated by the same frequency.

10.3 Stress sensors

10.3.1 EXTENSIOMETRIC GAUGES

10.3.1.1 General points

Extensiometric gauges are stress sensors, usually used to measure the deformation of a mechanical structure in order to calculate the pressure being exerted on it.

They are often used in the aeronautics industry, which has contributed largely to their widespread use. In robot technology, the use of this sensor has become more common in stress servocontrol, for example.

10.3.1.2 Method of use

The gauge takes the form of a very fine electrical conductor on a support, as shown in Figure 10.11. The snake-like shape allows most of the conductor to be oriented in a preferred direction. When the gauge is used, the deformations cause variation in its resistance, which can be measured.

If ℓ is the length of the conductor, D the diameter of its section and ρ the specific resistance of the material, then the resistance R of the

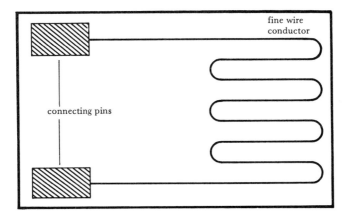

Figure 10.11. *Strain gauge*

conductor can be expressed as:

$$R = \frac{4}{\pi} \rho \left(\frac{\ell}{D^2}\right) \quad (10\text{-}1)$$

During the process of deformation, the variations in length and diameter cause a variation in resistance, which can be calculated by taking the logarithmic derivative of the expression:

$$\frac{dR}{R} = \frac{d\rho}{\rho} + \frac{d\ell}{\ell} - 2\left(\frac{dD}{D}\right) \quad (10\text{-}2)$$

Using the Poisson module μ, which links the relative variation of the diameter with that of the length,

$$\frac{\Delta D}{D} = -\mu\left(\frac{\Delta \ell}{\ell}\right) \quad (10\text{-}3)$$

the following is obtained:

$$\frac{\Delta R}{R} = \frac{\Delta \rho}{\rho} + (1 + 2\mu)\frac{\Delta \ell}{\ell} \quad (10\text{-}4)$$

By making the reasonable assumption that there is no variation in ρ, the important conclusion that $\Delta R/R$ is proportional to $\Delta \ell/\ell$ is obtained:

$$\frac{\Delta R}{R} = C\left(\frac{\Delta \ell}{\ell}\right) \quad (10\text{-}5)$$

Coefficient C is determined experimentally. It is a function of the metal used, and varies widely from case to case, as can be seen from the following values:

- C is 0.5 for manganin (copper, nickel, manganese);
- C is 2 for constantan;
- C is 2.2 for nickel-chrome;
- C is 3.5 for elinvar.

This property is employed when measuring deformations.

10.3.1.3 Principles of extensiometric measurement

The method is based on the use of the Wheatstone bridge (see Figure 10.12), where:

$$\frac{R_1}{R_2} = \frac{R_4}{R_3} \tag{10-6}$$

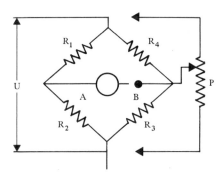

Figure 10.12. *Wheatstone bridge, with balance potentiometer*

In practice, the four resistances (R_1, R_2, R_3 and R_4) in Figure 10.12 often take the form of gauges. The advantage of the complete device is that it acts as a linear device. Some other systems are non-linear, notably those with only one gauge, the Wheatstone bridge not having a linear output as a function of the variation of just one of its resistances. Thus, gauges are usually positioned so that the parasitic effects can be cut out, and cancel out large stresses, increasing the sensitivity of the measurement.

10.3.1.4 Uses

Detailed information on the use of these sensors should be sought in an appropriate reference work, but the following precautions should be observed: the voltage generator used must have a high degree of stability, and the connecting wiring system should be devised with care, since errors from the resistance of the wiring can arise as well as variation of resistance with temperature. Signals produced by the gauges are weak and must be amplified. Generally amplification is in two stages, each amplifier having high-input impedance and low-thermal drift.

There are extensiometric gauge conditioners available on the market. The advantage of these gauges lies in their d.c. applications.

10.3.2 PIEZOELECTRIC SENSORS

Piezoelectricity uses the property of certain crystals, which when put under pressure generate electric charges on two opposite faces proportional to the effort applied. By carefully cutting the crystal, usually quartz, multidirectional sensors are obtained. These sensors do not easily operate with a d.c. supply.

Sensors of this type can be sensitive both to a force along an axis and a torque about the same axis.

10.4 Acceleration sensors

Acceleration sensors are used in the dynamic control of industrial robots. Measurement can be made in a number of ways:

1. It can be deduced by speed measurements. This is rarely fully satisfactory because of diminution in the signal-to-noise ratio.

2. The force created by the acceleration of a known mass can be measured. Such sensors use strain gauges. Then:

$$\frac{\Delta R}{R} = C \frac{\Delta \ell}{\ell} \qquad (10\text{-}7)$$

If a force F is exerted upon surface S of a material with Young modulus E, then by Hooke's law:

$$\frac{F}{SE} = \frac{\Delta \ell}{\ell} \qquad (10\text{-}8)$$

Consequently:

$$\frac{\Delta R}{R} = \frac{KF}{SE} \qquad (10\text{-}9)$$

3. A force, ie an acceleration equal to that which is to be measured, can be produced in a known mass, so that the mass remains in equilibrium. This force can be electromagnetic or electro-dynamic in origin and the equation is reduced to provide a measure of current. *Servo-return sensors* function on this principle, and are by far the most accurate.

It is difficult to measure angular acceleration. One method is to measure the tangential acceleration at a point situated at a fixed distance R from the axis of rotation, the position θ of which is noted. The resulting value of $Rd^2\theta/dt^2$ is found, and thus the required angular acceleration $d^2\theta/dt^2$.

The main parameters affecting the choice of a sensor are range of movement, range of measurement expressed in g (gravitational acceleration) and accuracy. Acceleration ranging from tenths of a g to thousands of a g can be measured. Models are also available having frequency band widths ranging from a few hertz to several kilohertz.

Chapter 11
External sensors

The vast majority of existing robots are not capable of external sensing but it is likely that by 1990 25 per cent of robots will be equipped with artificial vision systems.

11.1 Applications of external sensors

A robot is designed to perform tasks (ie to carry out physical work). The details and procedures involved in carrying out a task are described to the robot using an oral, written or gestural language that it can understand. To carry out a task, the robot must have access to information on predetermined parameters of the environment. The function of external sensors is to provide information about the environment and make these parameters available to the robot. The nature of these parameters depends on the tasks to be performed, and the information given to the robot during task description.

It is impossible to specify the external sensor without possessing a knowledge of both the task description and the task to be performed, all the more so because when a robot is constructed, the precise application is rarely, if ever, specified. To simplify this problem, the analogy of man and robot can be used: man is like a highly adaptable robot, and thus forms the ideal model. The functions performed by the senses in man are similar to those required of external sensors in robots (eg vision, touch, detection of stress, hearing).

The most complex task entrusted to the robot is that of handling and assembly. To carry out this type of task, the most useful sensors are those which provide *visual* and *tactile* information, and measure the levels of stress throughout operation, and for this reason these are the categories of sensor into which most research has been carried out. In 1979, for example, there were no robot vision systems available on the market, in 1982 there were 11 (4 Japanese and 7 American), and by the end of 1983 there will probably be more than 25 systems available.

11.2 Tactile sensors

11.2.1 ISOLATED BINARY CONTACT SENSORS

The technology used can vary, but the basic system consists of a switch with two positions, open and closed. Its position on the body of the robot is a matter of importance. When placed on the robot body (on the arm), its only function is to ensure safety in relation to obstacles, but when placed on the end effector, for example a gripper, it can provide more strategic information. This type of sensor is produced by a number of manufacturers, and can provide levels of repeatability of 1 μm and resolution of 2 μm.

11.2.2 INDIVIDUAL ANALOG SENSORS

In principle, the individual analog sensor is a flexible system with output proportional to local stress. It can be used for detecting position or stress. Depending on its position on the robot body, and its association with other sensors, its function can vary, as can the technology on which it is based. Its positioning is generally determined by the individual circumstance, and Figure 11.1 shows three very different examples of this.

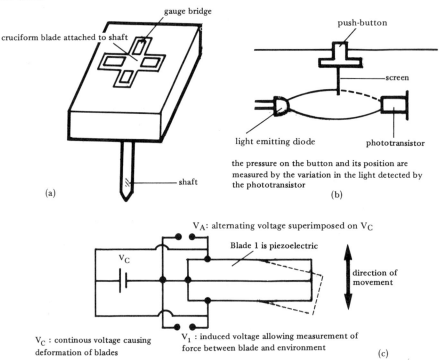

Figure 11.1. *Three examples of individual analog sensors: (a) sensor gauge bridge allows contact between the end of the shaft and the environment; (b) photomechanical transducer; (c) micromanipulator finger with tactile sensor*

11.2.3 MATRIX SENSORS

Matrix sensors are formed by combining simple binary or analog sensors in the form of a matrix. Each simple sensor is identified by the intersection of the row and column where it is positioned. If each individual sensor can provide information concerning force or position, the matrix grid provides, by integrating the simple information, complex data on the shape of an object. The analytical information technique is called *form recognition*, a topic too complex for inclusion in this book.

Figure 11.2 shows examples of the technology used in such systems, although as yet none has been perfected.

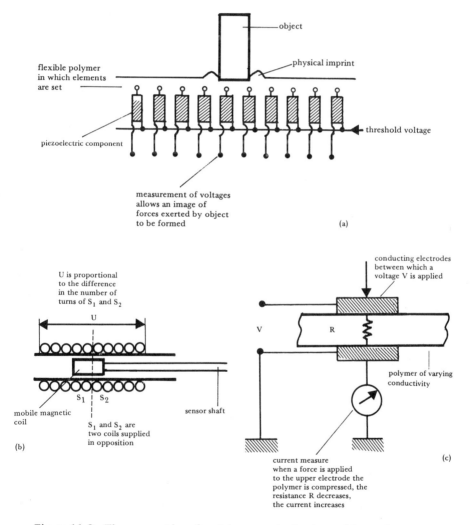

Figure 11.2. *Three examples of matrix sensor technology: (a) matrix sensor with a piezoelectric element; (b) matrix sensor element with needle effecting differential transformer; (c) artificial skin*

11.3 Stress sensors

In Chapter 10 on internal sensors, mention was made of strain gauges, which are the most sensitive part of stress sensors. In robot-environment interactions, it is important to *detect, locate* and *characterize* the reaction forces so that this information can be used in the task execution strategy.

Consider, as an example, the case shown in Figure 11.3, in which the force F, exerted between surface T and the object held by the robot gripper D, is to be found.

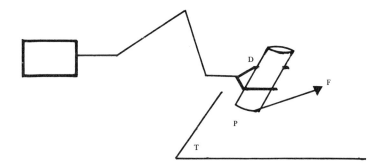

Figure 11.3. *Object-surface interaction*

There are three main methods by which this can be done:

1. The environment can be equipped with sensors: The object may be, for example, in contact with an instrumented platform. These platforms are in the form of plates of varying thickness, between which are arranged gauge bridges detecting forces in specific directions. These allow the coordinates of the point of contact and the force developed there to be found (see Figure 11.4). The arrows show the direction of the forces detected by the gauges. Thus, if a force is exerted on P, the coordinates of this point can be found as well as the movements of F relative to OX, OY, OZ and the components of F along the three axes.

2. The robot wrist is equipped with instruments (see Figure 11.5): This structure is based on the same principle as the instrumented platform, but it is adapted for assembly on the end segment of the robot. In this case also, the wrist allows the three movements and three components of F (see Figure 11.5) along the axes of an associated coordinate set to be found.

3. Couple modification as the actuators are used: If the robot is reversible, that is if a force F exerted on the gripper is 'felt' by the motors, the variation in the torque of the motors can be used to find the characteristics of the reaction force.

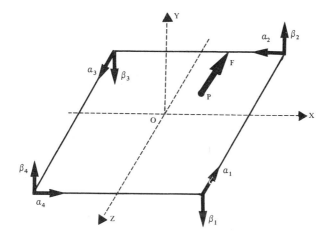

Figure 11.4. *Diagram of an instrumentated platform*

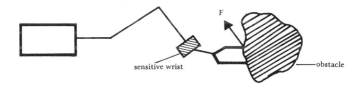

Figure 11.5. *Robot with a sensitive wrist for detecting obstacles*

11.4 Proximity sensors

These sensors are used to detect the presence of an object in the robot work space, and also to estimate the distance between the sensor and the object. The main methods used involve ultrasonics and infra-red techniques. A technique that uses air jets is still at the experimental stage.

11.4.1 ULTRASONIC SENSORS

Ultrasonic sensors can be used as presence detectors and for measuring distances. These sensors measure the time between emission and reception of ultrasonic waves from the object. Such sensors cannot be used for distances less than 30 to 50 cm, and their range is large. They are most commonly used in mobile robots to avoid obstacles. They are occasionally used experimentally in the grippers of large robots.

11.4.2 INFRA-RED SENSORS

An emitter (often an infra-red diode) sends an infra-red beam towards the object, which reflects and returns it to a receiver (often a phototransistor). To counteract disturbance effects from the ambient light, the light emission is modulated or pulsed (a few kilohertz) and filtered when received (see Figure 11.6).

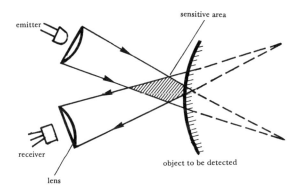

Figure 11.6. *Principle of an infra-red proximity sensor*

The advantage of these sensors is that they are all small (the emitter and receiver take up only a few cubic centimetres), and therefore can be used in robot grippers.

Although these sensors detect the presence or absence of an object in the work space easily, their application for measuring distance is more complex, since the light reflected by the object and returned to the receiver changes with the nature of the object (which absorbs light to a greater or lesser extent) and the orientation of its surface relative to the axis of the sensor. Moreover, if such a sensor is perpendicular to a flat surface, the response passes through a maximum value, which shows that, for a given value of response, there could be two possible distances (see Figure 11.7).

11.5 Visual sensors

Sight is one of the richest senses in that it allows a considerable range of information to be derived from a scene. This is reflected in the complexity of perception devices in artificial vision systems and the sophistication of the data-processing involved.

Thus, whatever the method used, a robot artificial vision system will always comprise:

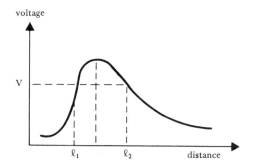

Figure 11.7. *For a value V of infra-red sensor response, there are two distances between the object and the sensor*

1. one or more signal emitters, either natural (eg reflection of ambient light by the objects in question) or artificial (eg flashes, lasers);
2. one or more sensors to receive these modified signals (eg cameras where the images produced are said to be primary, but they need not be necessarily optical and can be ultrasonic);
3. conversion of these images so as to be acceptable to the machines which will process them later (eg conversion of optical signals into electrical signals is the most common transformation);
4. any faults in the images are corrected, cleaned etc;
5. *image representation* when the above images are transformed to provide the necessary information;
6. *feature extraction*, in which *pertinent information* is derived from the results of various laws, algorithms and other criteria;
7. *recognition stage* when these characteristics are compared with the characteristics recorded during a training phase. Recognition can be total, partial or nil. Whatever the result, the robot must make a decision on the action to be followed as a result of the recognition process. Uncertainty in performance can arise due to errors at this stage.

One aspect of recognition still not completely understood is presenting the image in three dimensions. Three-dimensional vision systems are those in which the sensors provide directly an image in relief. Two-dimensional systems are those which provide a flat image.

Sensors used in vision systems can be diverse, and are the subject of much research and study, particularly with regard to three-dimensional systems. In practice, standard television cameras are most commonly used because of the low cost involved, but *charge coupled device (CCD)* cameras are used also, because of their reduced volume (ie they can be integrated into the structure of the robot) and their capacity for spatial discretization.

11.5.1 VIDICON-TYPE TUBES

The Vidicon is the most commonly used vacuum tube. It has an electron gun which sweeps a photoconducting mosaic on a transparent metallic film at constant speed. The light incident to the film passes through it and strikes the mosaic, changing its electrical resistance. When the beam of electrons strikes a point in the mosaic, the current received by the metallic film depends on the resistance at the point, related to the energy of the light which strikes it, and to the time interval between two successive sweeps of the electron gun. To establish a signal which is related to light energy only, the time interval between two sweeps of the electron gun must be constant, whatever the point on the mosaic.

The nature of the mosaic has a considerable influence on the characteristics, and there are different types of tube, differentiated by choice of mosaic.

11.5.2 CHARGE-COUPLED DEVICE CAMERAS

These are sensors which are said to be 'solid', as opposed to the vacuum tubes mentioned above, which are fragile. These systems are slightly more shockproof and have the advantages of decreased weight and volume, and compatibility with the processing systems.

These sensors take the form of *arrays* (between 256 and 2048 points) or *matrices* (up to 300^2 or 500^2 points) of photosensitive elements. These elements can be phototransistors or photodiodes, depending on the technology used (this makes no difference to the user).

The electrical signals are weak, and are amplified using amplifiers integrated into the sensor.

Chapter 12

Computer control

The actuators and sensors in the robot are driven by a control device, which can be, in the case of the sequence robot, a simple magnetic recording device or a programmable automaton, or a computer or computers, ranging from a microprocessor to a large and complex computer. The current trend is on one hand, moving towards decentralized control and the use of a microprocessor per function or per axis, and on the other hand, controlling the entire unit by a single, central microprocessing unit, with large memory storage. A summary of microprocessor operation in relation to robot control will be provided in this chapter.

Consider a robot with DOF θ_1 and θ_2 (see Figure 12.1). What elements are required to place the robot in the configuration $\theta_1 = \theta_1^d$ $\theta_2 = \theta_2^d$?

Figure 12.1 shows the two servo-systems, which are assumed to be equal, with motors M1 and M2, which drive the articulations θ_1 and θ_2.

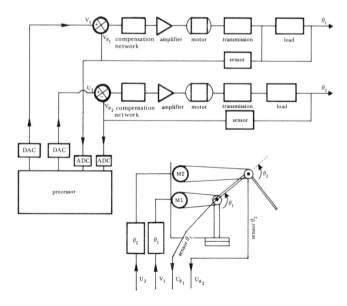

Figure 12.1. *Two servo-systems, which are assumed to be equal, with motors M1 and M2, which drive the articulations θ_1 and θ_2*

To place θ_1 and θ_2 in the required positions, the values of these two angles must be measured first, and control signals V_1 and V_2 derived.

A problem immediately arises concerning the nature of the signals V_1 and V_2. $V\theta_1$ and $V\theta_2$ are voltages or analog currents, whereas the processor can only receive digital signals. Thus, it is necessary to add an analog-digital converter to the processor, and to ensure that the output is connected to the servocontrol system of a digital-analog converter.

12.1 Analog-digital, digital-analog converters

The analog-digital converter (ADC) transforms a voltage into its binary representation and the digital-analog converter performs the reverse operation. These converters take the form of circuits using various methods of conversion. The main characteristic of an ADC are its definition (ie the number of bits resulting from the conversion); its frequency of conversion (ie the time between the command to convert and the moment at which the stabilized linear result can be used) and its accuracy and resolution.

Some converters are said to be *microprocessor compatible*. As well as the conversion circuits themselves, they have circuits that favour the acquisition of the binary value from the computer.

12.2 Other types of converter used in robotics

The circuits mentioned above are standard. For some applications they can be replaced by specific circuits, as in the following two examples.

12.2.1 SYNCHRO/RESOLVER: DIGITAL CONVERTERS

The synchro/resolver was mentioned in Chapter 10, where it was discussed in its more common role of angular sensor in motor servosystems. There are also specialized circuits that convert the angle into a binary code, which can be as large as 16 or 18 bits (resolution 5 seconds of arc). It should be noted that these circuits provide information concerning speed on an ancillary output, which avoids the need for tachometric generators.

12.2.2 VOLTAGE-FREQUENCY CONVERTERS

Although conversion takes longer than with a standard converter, these voltage-frequency systems are less expensive and have the advantage of the clock in the microprocessor.

To return to the example of the robot with two DOF, the decision of whether to retain the signals in the processor or transmit them to the servo-motors can be deferred.

The problem of telling the processor to receive or send signals is overcome by using a program.

12.3 The program

The nature of the program depends on knowing what is expected of it and involves its mode of operation.

12.3.1 KNOWING WHAT IS EXPECTED OF IT

This refers to the program in the general sense, that is to the algorithm the robot is to execute. In the example we are considering the program might say:

1. To go to $\theta_1 = \theta_1^d$ and $\theta_2 = \theta_2^d$
2. Calculate $V_1(\theta_1^d)$ and $V_2(\theta_2^d)$
3. Measure $V\theta_1$ and $V\theta_2$
4. Calculate θ_1 and θ_2 and store them
5. Calculate $\Delta V_1 = V_1(\theta_1^d) - V\theta_1$ and $\Delta V_2 = V_2(\theta_2^d) - V\theta_2$
6. Send $\Delta V_1/2$ and $\Delta V_2/2$ to servo-systems
7. If $\Delta V_1 < \epsilon$ and $\Delta V_2 < \epsilon$ stop
8. go to 3

The first comment about this program is that instructions 2 and 4 involve calculations of values external to the program and thus must be memorized somewhere. Instruction 4 concerns the model of the sensor and instruction 2 concerns the model of the robot itself. Moreover, in instruction 1, the task expected of the robot is described, and in instructions 5 to 8, the mode of execution of this task is described. Very complex tasks require special-purpose languages for task and robot description.

12.3.2 MODE OF OPERATION

The microprocessor performs its functions under quartz clock control. The frequency of the clock varies, according to the model, approximately between 2 and 10 megahertz. The surveillance and development of robot control thus require a certain amount of time. If too many operations are made, there is the risk of not being able to obtain controlled movements of the robot as rapidly as desired. In the simple example given above, this risk does not occur, but in more complex robots with vision systems, controlled in dynamic mode, the processing is so complicated that a single microprocessor would be unable to cope with either the capacity or the speed involved.

It is easy to forget how lengthy the calculations made in the course of a program are. This is especially true for trigonometric or transcendental functions (10 milliseconds for a sine arc would not be uncommon), but it is precisely these calculations that are required for a robot model (see Chapters 6 or 7).

To improve this situation, the following steps are carried out:

1. tabulation of functions (with interpolation, if necessary, between two neighbouring, tabulated values);
2. calculation of operators, which are specialized systems operating more or less in parallel with the main calculation, and only affecting it at the beginning and end of the process;
3. wired logic, by which the operation can be carried out very rapidly, no matter its complexity. However, only some operations can be wired in this way.

12.3.2.1 The language

The processor understands certain languages, with rules that must be respected. The languages available are at two levels:

1. the assembly language, which assumes a knowledge of the processor structure;
2. evolved or high-level languages, which are present or not (ie the microprocessor either has or does not have a translation program for the language, known as the compiler). Almost all microprocessors have an evolved language called BASIC. To this, FORTRAN or PASCAL, also general languages, can be added. It is important to remember that the compiler of each language reduces the memory space available for programming.

Languages specifically intended for robot programming are needed, and there are a considerable number available (eg AML, VAL, AL, SIGLA, RAPT, ROL, LM, LAMA), but so far none has become universally adapted because of various imperfections and the problems of adaptation from robot to robot, which prevent their becoming generally viable.

12.3.2.2 The program itself

The program is a succession of instructions written in a language understood by the processor, which is performed in sequence by the central processing unit. Depending on the level of the language, one instruction can activate many operations.

In the assembler language for a microprocessor, there are a number of main categories:

1. arithmetical instructions (eg addition);
2. logical instructions (eg and, or);
3. instructions of comparison;
4. loading and storing instructions (in the registers);
5. incrementation instructions;
6. jump instructions (conditional or otherwise);
7. input-output instructions and others.

There is another possible form of robot control, which is related to the presence of the interrupt procedure in microprocessors (for which the user must devise a program, which is difficult without facilities).

When a microprocessor controls a physical task, information necessary for the ensuing movements might have to be derived from the process (eg the signal from a strain gauge during contact between the robot and the environment). This signal can interrupt the central processing unit in the following way:

1. the central processing unit carries out the instruction in sequence;
2. it stores the information that will enable it to resume execution of the program after the interrupt in a specialized memory;
3. it reads the address of the sub-program to be executed in response to the interrupt;
4. it executes the sub-program (the last instruction tells it to resume the interrupted program);
5. it resumes execution of the interrupted program.

12.3.2.3 Structure of microcomputers

It was explained in Section 12.3.2.2 that low-level languages used by microprocessors include operations such as loading and storing. These are operations affecting binary values that are placed in precise locations.

The microcomputer (see Figure 12.2) has three major circuits: the *central processing unit*, the *memory circuits* and the *input-output circuits*. These three sets are connected by conductors, known as *buses*, which transport three types of values in the form of binary information. These communicate:

1. the data or information to be processed;
2. the addresses or positions of these data in the memories;
3. the control words in the input-output circuits.

1. The central processing unit carries out the essential operations on the data using an arithmetical and logical unit. Before entering this

unit, the data are stored in an *accumulator*, and before that, in *temporary storage registers*. Data are stored originally in the *memory circuits*.

2. There are two types of memory circuit:

— *Dynamic random access memory (RAM)*: Data can be written and read at will, with each new entry erasing the one before. This type of memory is used for modifying temporary results and information.

— *Static read only memory (ROM)*: Data can only be read by the central processing unit. The information is retained even if the electrical supply is cut. EPROM can be erased by ultraviolet light and reprogrammed using special equipment. EEPROM can be erased and reprogrammed electrically from the central processing unit.

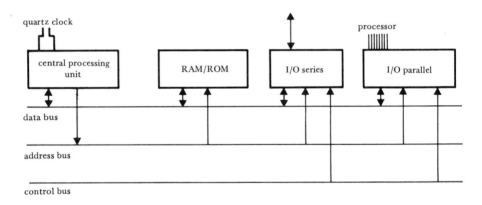

Figure 12.2. *Structure of a microcomputer*

3. The input-output circuits can vary, but there are two basic types of connection: *series* and *parallel*. In the series mode, data are transferred byte by byte. These bytes are separated by the control bits (the circuit is initialized first). In the parallel mode, the byte is transmitted (or received) in eight distinct entries.

The robot control program is in the assembler, and very similar to the structure in which the different components have addresses given by the compiler.

12.4 Conclusions

In this chapter, attention has been drawn to some of the rules governing the choice of microprocessor for robot control. These considerations can be summarized as follows:

1. the nature and number of the inputs and outputs;

2. the choice of converters;
3. the basic speed of the microcomputer and the number of bits it can handle;
4. the capacity for wired or specialized operations;
5. the dynamic memory size;
6. the use of languages and space for the compiler;
7. the handling of interrupt.

Chapter 13
Robot training and trajectory generation

Strictly speaking, a robot must go through four major stages when performing a task; it must:

1. understand what is required of it;
2. be aware of its starting position;
3. generate a strategy for performing the task;
4. carry out this strategy.

In industrial robots commonly in use, these four stages are not strictly observed. The robot is taught a number of trajectories, and it is activated by a program and performs the appropriate move. The robot does not understand the problem, nor does it analyse its starting situation. All is controlled by sequential programming; there is little or no strategy (except conditional orders: 'If the limit sensor is activated, do this or that'). There is no real strategy applied.

13.1 Methods of recording trajectories

There are two possible methods for recording trajectories. They depend on whether the robot is trained passively by an operator (see Figure 13.1), or whether it is manually controlled from a master station (see Figure 13.2). In the first case, successive values of the articulated variables, provided by the sensors, are recorded. This information is then translated into control voltages for the servo-system before the robot functions automatically. This can be done by analog or digital

Figure 13.1. *Passive training by operator*

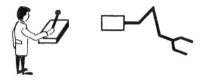

Figure 13.2. *Manual control from a master station*

processing. In the second case, servocontrol voltages are used directly during the training procedure and the robot is therefore *active*. Only the second procedure can be used for direct recording.

13.1.1 PERMANENT CONTINUOUS RECORDING

The servocontrol input voltages to the motors during manual operation can be recorded on magnetic tape (eg for a robot manipulator possessing six DOF and an end effector gripper, seven tracks are recorded, unless multiplex is used). By returning to the initial configuration after task execution, the manoeuvre can be repeated in the automatic mode, by replaying the recording, with the correct gain at the recorder output. A digital computer is not suitable for this.

13.1.2 PERMANENT DISCRETE RECORDING

The same type of recording can be obtained using a digital computer with a large memory capacity. During execution, the reverse operation occurs by way of an initialization process (see Figures 13.3 and 13.4).

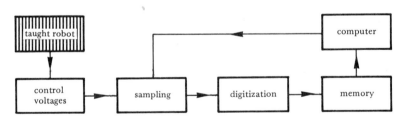

Figure 13.3. *Digital recording using an active robot*

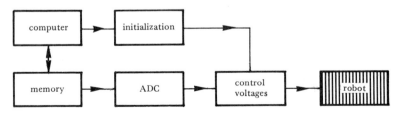

Figure 13.4. *Automatic execution (training with an active or passive robot)*

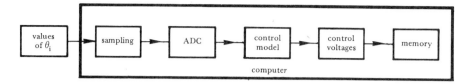

Figure 13.5. *Recording with a passive robot*

When the robot is passive during the training process, the data obtained from the sensors must be translated into control voltages (see Figure 13.5).

13.1.3 RECORDING SPECIFIC CONFIGURATIONS

To manoeuvre a robot by hand, either directly or through the intermediary of a master station, is not always easy. Problems can arise if the desired manoeuvre is to be performed more rapidly in the automatic mode than during the training process. Moreover, it is easy for the robot to make inaccurate movements, for instance by overshooting or oscillating. The faults can be overcome in the following way. The task to be performed can be divided into a number of stages, each characterized

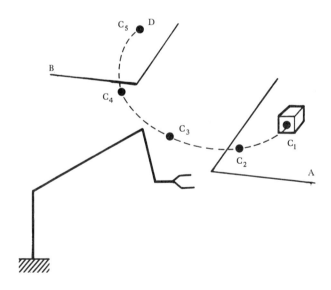

Figure 13.6. *The robot must take the object which is on plane A, and move it to D, on plane B. It is led successively through the configurations corresponding to points C_1, C_2, C_3, C_4 and C_5*

by a starting configuration and a final configuration, which is the starting configuration of the following stage. The robot is moved manually into each of these configurations, which are known as the *important configurations*. These and only these are recorded. An example is given in Figure 13.6. During automatic task performance (or during the training phase) the computer will generate a trajectory for moving from one important configuration to the next.

13.2 Manual control used in training

13.2.1 MANUAL MOVEMENT OF THE ROBOT

This procedure is used in applications such as paint-spraying and consists of the operator steering the robot through the routine. The trajectory (ie the angular values of the articulations) is recorded and can be reproduced in the automatic mode as many times as required. The starting command is obtained from a synchronized relay which is triggered by the objects to be sprayed as they move along a conveyor belt. In this circumstance, the robot should not present too great a resistance to handling.

13.2.2 USE OF A STEERING MECHANISM (SM)

The *steering mechanism* (SM) is an AMS controlled by an operator. It links the movements of the operator with those of the robot using an electronic, computerized or mechanical coordinate changer. The SM can take various forms, depending on the function of the robot. Here follows a description of three basic types available.

13.2.2.1 Master arm

The master arm is used in master-slave teleoperation systems. The structure of the SM is identical to that of an ordinary robot except that the gripper is replaced by a handle (see Figure 13.7). This type of system is used in telemanipulation as it leads to a symmetry between the master

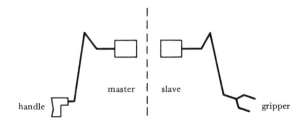

Figure 13.7. *Master-slave couple*

arm and the slave arm, and so handling is easier. If the master arm is pushed forward, the slave arm moves forward accordingly. The gain between a rotation of the master articulation and the corresponding rotation of the slave arm can be varied. In some models, positional shift is possible, so that the slave is moved relative to its original position. This system is used particularly in nuclear power stations, foundries and for underwater exploration etc.

13.2.2.2 Manipulator

The manipulator is used in such applications as paint-spraying. This is a light structure, often having the same number of DOF as the robot, which is easily manoeuvred. The task is first performed with the help of the manipulator, and the robot learns the trajectories which it subsequently reproduces.

13.2.2.3 Joystick

The *joystick* has the same number of DOF as the robot. The operator can guide the end effector using a joystick. When the joystick is moved forward, the gripper advances; if the joystick is pulled back, the gripper moves back; if the joystick is moved to the right or left, the gripper moves accordingly; if the joystick is pulled up or pressed down, the gripper rises or descends (see Figure 13.8). Hence, manipulations in any direction can be achieved using a joystick.

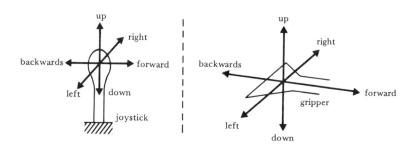

Figure 13.8. *Joystick control*

Some joysticks are equipped with pressure control; all the movements mentioned above can be achieved without moving the joystick, but by applying pressure in the appropriate direction.

In all these sequencing mechanisms, the principle is the same: movements of the master device must be translated into movements of the robot. Importance is attached to the *ergonomics* or in other words how efficiently the master translates the operator's instructions to the slave.

13.2.3 PUSH-BUTTON CONTROL

This is the oldest type of control. In this system, each DOF of the robot is connected to a switch. The operator must decide on the best transformation using the reasoning: 'If I press button 3, angle θ_3 increases and the gripper moves forward etc'. Some manipulators have mixed control: the carrier arm is controlled by a joystick, and the end effector is oriented using a system of push-buttons. In more elaborate systems, the switches correspond to specific commands, such as 'move the gripper down, up, left, right etc'. In some cases, the more pressure applied to the switch, the faster the robot moves (*velocity control*).

13.3 Trajectory generation

If, during training, two successive important configurations C_1 and C_2 are recorded, how is the trajectory which the robot will follow from C_1 to C_2 produced?

First solution: In this method the time factor, which dictates the moment at which C_2 should be reached, is ignored. This requires dynamic control, the difficulties of which have been shown already.

Second solution: Since the important configurations already have been chosen, it will be assumed that there are no spatial constraints (ie no obstacles between C_1 and C_2) so any trajectory can be made. The choice is infinite.

Third solution: It is assumed that there are no constraints linked to the performance of the task, or affecting the robot. For example, if a vessel containing liquid is to be moved, the container must remain vertical throughout.

These three solutions can be tried, one after the other, starting with the simplest example.

The configurations C_1 and C_2 are known by the values of the articulated variables. For example, for a robot with six DOF:

$$C_1(\underline{\Theta_1}) = (\theta_1^d, \theta_2^d, \theta_6^d) \qquad (13\text{-}1)$$

$$C_2(\underline{\Theta_2}) = (\theta_1^a, \theta_2^a, \theta_6^a) \qquad (13\text{-}2)$$

First an increment $\Delta\theta_{i\,min}^{\,max}$ is fixed, which represents the smallest of the individual maximum increments of the angles θ_i carried out during a sampling period T of the control. For example, if kinematic control is to be used, during the control period T, θ_1 can be moved by a maximum of 5 degrees, θ_2 by 4 degrees, θ_3 by 6 degrees, θ_4 by 2 degrees, θ_5 by 8 degrees, θ_6 by 8 degrees so that the model remains valid. These are the individual maximum increments. The smallest of these increments is chosen: the 2 degrees of θ_4, which is called δ.

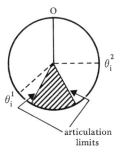

Figure 13.9. *To move from θ_i^1 to θ_i^2, it is necessary to pass through zero*

In the robots shown in Figures 3.2 and 3.4, it is easy to see that, taking into account the lever arms, the masses and inertias, it is possible to vary θ_5 for example, more than θ_2 during the same time interval T, without endangering the validity of the kinematic model.

Using δ, all excursions of θ_i, going from C_1 to C_2 can be measured. These excursions must be compatible with the articulation inputs, so the directions must be checked. For example, Figure 13.9 shows that when moving from θ_i^1 to θ_i^2 it is necessary to pass through zero.

If this precaution is taken, the following can be written:

$$\theta_1^a - \theta_1^d = N_1 \delta$$
$$\vdots \qquad \vdots \qquad (13\text{-}3)$$
$$\theta_6^a - \theta_6^d = N_6 \delta$$

The largest values of N_i are chosen. For example, N_4 would be the number of steps necessary to go from C_1 to C_2 for each variable θ_i. The angular variation for each value of θ_i can be performed using increments of amplitude:

$$\Delta\theta_1 = N_1 \delta | N_4$$
$$\Delta\theta_2 = N_2 \delta | N_4$$
$$\vdots \qquad (13\text{-}4)$$
$$\Delta\theta_4 = \delta$$
$$\vdots$$
$$\Delta\theta_6 = N_6 \delta | N_4$$

After calculation, the series of control vectors $\underline{V} = \underline{V_1} + K\Delta\underline{V_1}$ is recorded, which allows the movement to be performed automatically. It should be noted that in some systems (eg the robot IRb-6 manufactured by ASEA) t_1^2 can be imposed between two configurations. Thus, the system can calculate, not only the number of increments in θ_i, but also the sampling period T. If the time is too short, and thus incompatible with the computer's abilities, or with the kinematic solutions, the refusal is displayed, and t_1^2 must be increased to allow computation.

13.4 Trajectories in the task space and in the articulated variable space

The above calculation shows how to move from C_1 to C_2 in the articulated variable space. All that is certain is that the robot starts in configuration C_1 and finishes N_4 steps later at C_2. A trajectory in the $\underline{\Theta}$ space is chosen. If control vector $\underline{V}_1$ corresponds to the configuration C_1, and $\underline{V}_1 + \Delta \underline{V}_1$ corresponds to $\overline{\Delta \Theta_1^d}$, Table 13.1 can be written.

Successive configurations	Values of each articulated variable for each configuration	Control values of each actuator for each articulation	Successive control vectors
$\underline{\Theta}_1$ $\underline{\Theta}_1 + \Delta \underline{V}_1$	$\theta_1^d + N_1 \delta \vert N_4$ $\theta_2^d + N_2 \delta \vert N_4$ $\vdots$ $\theta_4^d + \delta$ $\vdots$ $\theta_6^d + N_6 \delta \vert N_4$	$v_1^1 + \Delta v_1^1$ $v_2^1 + \Delta v_2^1$ $\vdots$ $v_4^1 + \Delta v_4^1$ $\vdots$ $v_6^1 + \Delta v_6^1$	$\underline{V}_1$ $\underline{V}_1 + \Delta \underline{V}_1$
$\underline{\Theta}_1 + 2\Delta \underline{V}_1$	$\theta_1 + 2N_1 \delta \vert N_4$ $\vdots$	$v_1^1 + 2\Delta v_1^1$ $\vdots$	$\underline{V}_1 + 2\Delta \underline{V}_1$
$\vdots$	$\vdots$	$\vdots$	
$\underline{\Theta}_1 + N_4 \Delta \underline{V}_1 = \underline{\Theta}_2$	$\theta_1^d + N_1 \delta = \theta_1^a$ $\theta_2^d + N_2 \delta = \theta_1^a$ $\vdots$	$v_1 + N_4 \Delta v_1^1$ $v_2^1 + N_4 \Delta v_2^1$ $\vdots$	$\underline{V}_1 + N_4 \Delta \underline{V}_1 = \underline{V}_2$

Table 13.1

To find the important trajectory, $\underline{X}$ in the task space, the following equation must be applied at each step:

$$\underline{X}(R_0) = F(\underline{\Theta} - \underline{\Theta}_0) \qquad (6\text{-}33)$$

as seen in Chapter 6. If this trajectory is unsuitable, the inverse procedure should be used: for $C_1(\Theta_1)$ take $\underline{X}_1$ and for $C_2(\Theta_2)$ take $\underline{X}_2$. The interval $\underline{X}_2 - \underline{X}_1$ is divided into a number of steps $\underline{\Delta X}$. Applying the results of Chapter 7 and Appendix V the following is obtained:

$$\underline{\Delta \Theta} = [J]^{-1} \underline{\Delta X} \qquad (13\text{-}5)$$

Ensure that the $\underline{\Delta \Theta}$ are not too large. If they are, smaller increments $\underline{\Delta X}$ must be chosen.

13.5 Control languages

The high-level languages mentioned here have nothing in common with the programming examples given in this book. The latter are used for actuator and sensor control.

A high-level language is one in which the syntax is similar to the user's spoken language. There is a hierarchy in the degree of development of languages, and a classification which appears in *An Evolved Language for an Intelligent Robot*[1] is used here to grade the languages according to their complexity. The simplest language refers to notions of end effector movement. WAVE,[2] a language developed at Stanford, is the pioneer in this field. Others such as SIGLA (from Olivetti), EMILY (from IBM), LM (from Imag) and VAL (from Unimation)[3] have followed. VAL, which was perfected by a robot designer, is one of the few actually in use. In this language, the programmer uses the following types of command: 'Move towards point M at velocity V' or 'Move forward until the force component F on Z is >0.5'. These procedures often involve a long and exhaustive description of the task to be performed.

At a higher level, there are languages which directly describe the actions to be performed. In this system the programmer expresses the requirements in the following way: 'Insert the rod into the piston', that is, using a language very close to the task required. In this language programs would be quicker and written in fewer lines, since the user no longer has to describe the movement in detail. On the other hand, the computers required to handle this type of language are sophisticated. The program is often converted using another computer into a VAL-type program, or any inferior language, which is transmitted to another computer which controls the robot. AUTOPASS (from IBM) and LAMA (from MIT)[4,5] are examples of this level of language.

At the highest level the programs used are known as *plan generators*. These involve artificial intelligence, and will only be mentioned briefly, since they are not used in industry at present. These languages require highly complex programming. STRIPS developed at SRI and BUILD developed at MIT are languages of this type.

13.6 Conclusions

The problem of trajectory generation is large. This chapter claims only to present a very simple summary. Trajectory generation should not be confused with plan generation; the former is a part of the latter. Plan generation is synonymous with strategy generation. A problem presented to a robot with its plan generator chooses a solution for performing the task, which includes the choice and calculation of the set of trajectories it will employ.

References

1. Falek, D.; Parent, M. An evolved language for an intelligent robot. *International Seminar on Programming Methods and Languages for Industrial Robots* Institut de Recherche en Informatique et Automatique Roquencourt, June, 1979.
2. Paul, P. WAVE: A model based language for manipulator control. *The Industrial Robot* March, 1977.
3. Unimation. *User's Guide to VAL* A Robot Programming and Control System Version 11.
4. Lieberman, L.I.; Wesley, M.A. *AUTOPASS: An Automatic Programming System for Computer Controlled Mechanical Assembly* IBM Report RC5925.
5. Lozano-Perez, T.; Winston, P.H. *LAMA: A Language for Automatic Mechanical Assembly* 5th International Joint Conference on Artificial Intelligence Cambridge, Massachusetts, 1977.

Chapter 14
Robot performance and standards

14.1 What is robot performance?

When an industrial robot is to be designed, the potential user presents a load file. This load file defines the limits of use, and the performance of the robot can be measured in relation to these constraints. These might concern the weight and energy independence of a transportable machine, accuracy, speed of execution or reliability of another sort of device.

As stated earlier, once of the important qualities of a robot is versatility, in other words, the ability to perform various tasks not specified in advance. Since the use to which a robot will be put is not always known *a priori*, performance standards related to the specific parameters cannot be established, and attempting to enumerate any possible constraint and resulting performance of the robot is too complex to be practical.

In practice, the prospective user of the industrial robot should ask four basic questions concerning its practical performance:

1. Is the robot capable of performing the task or tasks required of it?
2. Does the use of the robot impose technical constraints (eg space, fluids, dust protection, strengthening the floor, link with production line and safety measures etc)?
3. What human resources (eg programming and maintenance etc) are required to run the robot?
4. What is the cost and cost-effectiveness (eg investment, upkeep and increase in productivity etc) of the operation?

If (2) and (4), which are cost-related, are grouped together, then analysis of robot performance can be considered in three parts.

14.2 Task performance

For a robot at work, and using its end effector, there are important practical characteristics to be considered.

14.2.1 ATTAINABLE VOLUME

This is expressed in units of volume, but the shape of the envelope, which can be complicated since it can be formed by a combination of several articulations, is important. It is simpler and more common to describe the attainable volume in terms of a surface, on which is written the real volume, but using a simplified shape, for example an intersection of spheres, parallelipipeds or ellipsoids. Otherwise, graphical representation is necessary. Hence, there are two parameters: volume and shape.

14.2.2 POSSIBLE ORIENTATION OF THE END EFFECTOR

Because of the limits of each articulation, and the mechanical couplings between articulations, it is not always possible for all three DOF of the end effector to attain their full range in the attainable space. Either the minimum angular range of the end effector relative to the fixed coordinate set for the whole attainable volume, or a map of this reduced range, could be provided. (A map is more useful than the information provided by manufacturers. It describes the reduction in range of each articulation relative to a reference point, whilst ignoring the problems of limits and coupling, and sometimes considerably modifying the system's actual capabilities.)

14.2.3 PAY LOAD

The couples to be developed by the actuators are a function of robot configuration. For this reason the maximum load the manufacturers suggest cannot always be transported everywhere in the attainable volume, or with all possible orientations. Two sets of information should be provided: the maximum load that can be lifted in the ideal configuration, and the effective load, which is the load that can be properly manoeuvred throughout the whole attainable volume. Alternatively, a graph showing configuration and pay load can be provided.

14.2.4 POSITIONAL PRECISION

There are three major aspects:
    1. If starting from any configuration the robot is required to move to another position, which is not close to the first, how accurately will the second configuration be attained? This is expressed by two values:

— precision of the terminal point of the arm or the centre of gravity of the end effector (in millimetres);
— angular precision of the end effector (in radians).

The first type of precision is sensitive to two parameters, for constant control:

- zone of attainable volume to be developed;
- transported load, since the articulation elasticity can be significant, and the inertia of the load modifies the dynamic behaviour.

For this type of static precision a guaranteed loaded value can be given, or a multidimensional map established.

2. If the robot is made to repeat the same movement many times, what is the repeatability?

3. Starting from a fixed position and orientation of the end effector, what is the smallest modification that can be made to these values?

The problem of resolution or sensitivity of the robot depends on the entire control and the technology available. If adjustments are to be made to the robot, for example for use in automatic assembly, then this parameter becomes very important. Sensitivity of the robot is also a function of the pay load and the configuration. A precise value can be guaranteed, but very few manufacturers do this.

If positional precision is a very important performance parameter, precision at speed (dynamic precision) is less so, and is moreover very difficult to measure.

14.2.5 VELOCITY

This is a fundamental characteristic of industrial robots. The time required to accomplish a task is of major interest to the user. In evaluation of a robot's performance *a priori*, the nature of these tasks is of course now known, thus only velocity of movement and rotation of the end effector can be provided. For example, during a control sampling period T, each articulation θ_i is varied by $\Delta\theta_i$. The positional and rotational movement of the end effector during this period is:

$$\underline{\Delta X} = [J(\underline{\Theta})]\underline{\Delta\Theta} \tag{14-1}$$

Thus, the instantaneous velocity in the task space is:

$$\underline{\dot{X}} \cong \frac{\underline{\Delta X}}{T} = [J(\underline{\Theta})]\frac{\underline{\Delta\Theta}}{T} \cong [J(\underline{\Theta})]\underline{\dot{\Theta}} \tag{14-2}$$

If $\underline{\dot{\Theta}}$ can have a constant value, $\underline{\dot{X}}$, which depends on the configuration $\underline{\Theta}$, cannot.

Most manufacturers are evasive on the subject of velocity control. Orders of magnitude and maximum velocities are only mentioned with reference to end effector translation. Sometimes maximum speed of articular rotation is mentioned, but it is difficult to find $\underline{\dot{X}}$ (or rather $|\underline{\dot{X}}|$), which is the value sought by the user.

14.2.6 RELIABILITY

Relating as they do to all the elements of a robot's performance, these particular data are of great interest to the user. This is why robot reliability is discussed under task performance.

As for any other systems, robot reliability is defined by its breakdown rate, which is expressed by a percentage time during which the robot cannot fulfil its assigned role. Reliability is often expressed using a frequency graph (frequency of breakdowns) for the whole life of the system. Robots are subject to two major types of breakdown:

1. total cessation in functioning;
2. deterioration in performance, for example in spatial precision: drift can arise and after a time the robot no longer can perform the task, or else a DOF can cease to function. In this case, it is useful to have a redundancy of the DOF, which allows the fault to be compensated for automatically until it can be repaired.

14.2.7 SYNCHRONIZATION WITH OTHER MACHINES

In manufacturing, successive tasks are temporarily linked. The robot must be capable of being placed in the production line with other machines and computers. The number of possible links (input and output) can be an important issue when evaluating performance.

14.3 Human performance in robot control

This is concerned with the ease with which the robot can be used on the shop floor and controlled by a novice. The skills required for controlling a robot are:

1. knowledge of the equipment: adjusting servo-systems etc;
2. knowledge of the software: programming the robot.

It is important to know whether a technician would be able to program and make adjustments to the robot, after a brief training period of a few days, for example, or whether an engineer would be needed. In the same way, would it be possible for the user to carry out continuous maintenance and minor repairs (for which diagnostic guides and test procedures could be provided) or would it be necessary for a specialist from the manufacturer to attend for all maintenance.

14.4 Economic performance

This concerns the economic viability of robot use in the factory, which will certainly improve in the future. It depends on a large number of factors:

1. investment costs:
   — purchase price,
   — installation cost,
   — cost of adapting workshop;
2. running costs:
   — fuel, maintenance, life expectancy, reliability (should an allowance be made for replacement by worker in case of breakdown?) and percentage duration of use;
3. nature of tasks to be performed:
   — are tasks expensive when performed by man?
   — will the production rate increase?
   — will the quality be improved?
   — will there be a reduction in workforce?
4. degree to which the workshop is already automated: placing a robot between two men or a man between two robots is only justified if the task is dangerous, or in other exceptional circumstances;
5. size of the business: a single robot may not be viable but ten used together might;
6. other factors.

The manufacturer can only offer robots with the 'best possible performance' for the 'best possible value', and hope to attract the most suitable users. The user should determine the standard of economic performance that is expected, after defining the tasks to be automated, and the human and task performances offered by the manufacturers.

14.5 Performance standards

Since robot performance cannot be defined clearly, it is extremely difficult to measure. After eliminating human and economic performance, the remaining parameters seem vague and difficult to fix. Only the task performance can be considered, and is reduced to measuring some of the factors of merit which are of major importance, such as positional precision. The most popular technique is based on the use of a cube (see Figure 14.1). The cube is fixed to the end of the robot at one of its faces. This is the male part which is to be fitted to a female cube from which three faces have been removed, transforming it into a trihedron. The trihedron is equipped with various sensors (proximity, optical and inductive etc) on its three faces which allow the position and orientation of the cube relative to the trihedron. The object of the exercise is to place the cube perfectly in the trihedron.

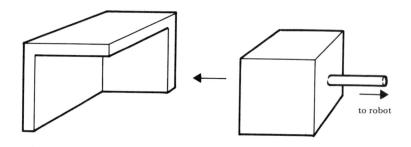

Figure 14.1. *Complementary cubes used for checking positional precision*

14.5.1 TECHNICAL EXAMPLES

The IPA (West Germany) system: The trihedron is equipped with three inductive sensors which measure small distances. The parameters to be varied are:

1. direction of movement (the cube is moved around the three possible axes of the trihedron successively);
2. load;
3. velocity;
4. position of the trihedron.

Static positioning is measured by sending the robot to its final configuration and measuring the error in three dimensions. Dynamic positioning is measured in the same way, but the robot performs a cycle, passing through the required configuration ten times. Inverting the direction of movement of the cube leads to another type of positioning error. Long-term repeatability is evaluated using a test cycle repeated for several hours. Finally, overshooting of a required position and damping of the resultant oscillations are measured.

The Renault system: For static measurement, a cube is used, on five sides of which are targets, and on the sixth side a target of diameter 0.5 mm. This target is lit by optical fibre, and the measurement is made using two theodolites.

The Peugeot system: The robot holds the female trihedron, with two proximity sensors on each side. Three test cubes are mounted on a mobile platform. Three wide potentiometers are used, with their extremities fixed to the end part of the robot, to check the trajectories. The trajectories can be followed by recording the potentiometric signals.

Chapter 15
Robots in use

15.1 Examples of uses

15.1.1 LOAD TRANSPORTATION

There are robots available that can transport loads ranging from a few kilograms to 1 tonne (eg the Andromat handling robot).

In the Renault factory at Cleon, the Renault ACMA robot is used for depalletization of 800 crankshafts onto a conveyor belt (see Figure 15.1). Each part weighs 12 kg. The operation lasts 45 minutes per pallet. Before the robot was introduced, three people were required. They handled between them several tonnes of metal a day. At Caterpillar, the palletization of track components (80 kg) is entirely performed by robot.

In hostile environments, such as foundries or forges, robots are often used for high-temperature handling, particularly in stamping operations. At Du-Wel Metal Products Incorporated a Unimate robot unloads two stamping presses of 800 tonnes. In the General Electric factory at Portsmouth six Unimate 2000 robots unload plastic television bodies from the presses.

15.1.2 FEEDING AND UNLOADING MACHINE TOOLS

Feeding and unloading of machine tools is an area in which robots are used, especially with automatic lathes. At the Caterpillar factory in Peoria, a T3 Cincinnati Milacron synchronizes the feeding and unloading of two lathes. The unmachined part is taken from a conveyor belt and placed in whichever lathe is free, for machining. At the same time a finished part is removed from the other lathe, taken to the laser verification point, then either placed on another conveyor belt or scrapped. This manufacturing unit can be supervised by a single person.

Figure 15.2 shows a similar operation with a robot made by Unimation. The end effector is a double gripper.

Figure 15.1. Use of the Renault ACMA robot for depalletization of crankshafts onto a conveyor belt

Figure 15.2. *Unimation robot with a double gripper end effector in use*

15.1.3 WELDING

Welding is an unpleasant and repetitive task, and one for which robots are already widely used, and are likely to be used more extensively, due to the progress made in the development of sensors. The technique of spot welding is well developed, and is used frequently, especially in the motor industry. Continuous arc welding is a far more delicate operation, the main problems being to follow the joint, and find a sensor which can function properly near a welding torch. The market for robots in continuous welding is optimistic. Some manufacturers offer robots specifically designed for this purpose. This is the case with the AW7 model by General Electric, the Westinghouse W7000 and the Linc-man by GKN Lincoln Electric Limited. Motor car manufacturers frequently use welding robots. Figure 15.3 (a and b) shows two robots on the car production line.

15.1.4 SMALL UNIT ASSEMBLY WORK

Robots are also widely used for assembly, either of printed circuit

Figure 15.3. *(a) Use of welding robots on the car production line (Renault ACMA)*

boards, into which components are automatically inserted, or of small systems.

In the first case, the required coverage for components has not been achieved, although active research is progressing in this field. At Carnegie Mellon University (The Robotics Institute) experiments have taken place, using several robots (1 Seiko, 2 Unimation Puma) and a visualization unit (a black-and-white General Electric camera). The system can organize the components to be inserted in advance: resistors, transistors and circuits. The whole process is controlled by a PDP 11/23.

In the second case, that of mechanical assembly, laboratory research led to industrial application. To assemble plastic connection terminals General Electric uses the Unimation Puma. IBM uses its own robot to assemble typewriter keyboards (inserting the touch pads and keys), and position the typing ribbon during impregnation.

In the SONY factory in Tokyo, there is a production line for their WALKMAN products. The parts for assembly are placed on a plastic surface equipped with protrusions which can position them, and small 'stores' where they can be piled up. The surface is divided into work areas, in which different stages of assembly, 48 in all, are carried out. The surfaces are placed on XY tables, which are positioned under the tools, with a precision of 15 μm. There are several points: oiling, gripping and joinging etc. The unit can produce 200,000 systems per month.

Figure 15.3.(b) *Use of welding robots on the car production line (Renault ACMA)*

15.1.5 PAINTING

The scale of use of robots in painting operations has not been spectacular, even though the task is unpleasant, and potentially dangerous. There are however several robots available which have been designed solely for this function. Renault uses its own Renault ACMA P7, which has seven DOF for painting vehicle bodywork. AKR (AOIP Kremlin Robotics), General Electric with the S6 model and others supply equipment for this use. Caterpillar uses a specially equipped TRALFA robot to paint its excavation machines. This firm attaches more importance to the automation of its painting processes than to arc welding at the present. Figure 15.4 shows a painting robot. The wrist guides the end effector during the training phase.

Figure 15.4. *A painting robot at work on the car production line*

15.1.6 SHOE CARDING

The final example is an unusual and recent use of robots: shoe carding. Shoe carding consists of removing the top layers of leather in a limited area, to reach lower layers, where the more resilient fibres allow better adhesion to the sole. This operation was automated with an ASEA IRb-6 robot, equipped with a special tool, with stress servocontrol, programmable according to the type of leather being used. Figure 15.5 shows this process.

Figure 15.5. *ASEA IRb-6 robot in the shoe-carding operation*

15.2 End effector components

In all the examples given in this chapter, the robots are equipped with end effectors specific to each application. These systems are often designed by the user, who best knows the unique characteristics of the product and can adapt the equipment accordingly. Much of a robot's usefulness depends on its end effector.

There are many types of gripper available for various types of manipulation. Each one can be equipped with force sensors or proximity sensors and can often be used for more than one activity. They can be driven electrically or hydraulically, or even by compressed air. Figure 15.6 shows a number of grippers suitable for use with different types of object: (a) tubes, (b) parts to be moulded and (c) composite materials etc.

In assembly, RCC systems (Remote Centre Compliance) are some-

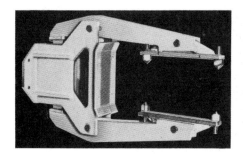

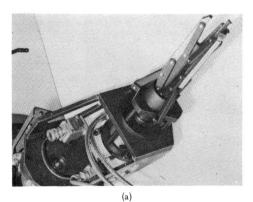
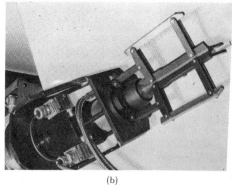

(a)   (b)

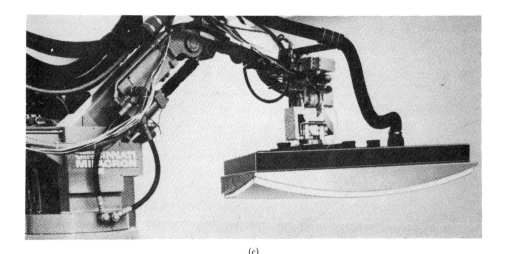

(c)

Figure 15.6. *Grippers used for different types of objects: (a) tubes, (b) parts to be moulded and (c) composite materials (Cincinnati Milacron)*

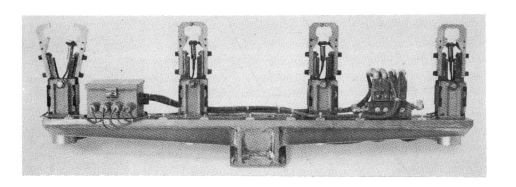

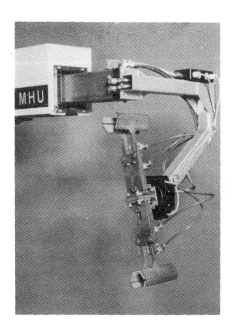

Figure 15.6. *(c) continued*

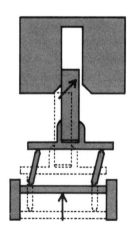

Figure 15.7. *Remote Centre Compliance (RCC) system. Such systems are mounted between the extremity of the robot and the gripper (see Figure 15.8)*

times used. These are mounted between the extremity of the robot and the gripper, and solve problems of poor adjustment in assembly operations. Figure 15.7 and 15.8 show the system and the principle.

Paint sprayguns that can be fixed onto the robot are now available. Figure 15.8 shows a system of this type, the Modan PPH 50X, which can paint 9 m² per minute.

Figure 15.8. *Paint spraygun to be fitted to the extremity of a robot (Modan)*

15.3 Conclusions

This survey does not pretend to take in more than a few of the most common uses to which robots are put. There are many others, and new developments are made every day. These developments often require new sensors, with better performance, or new tools, as well as the rethinking of production methods. The more specialized the task required of the robot (because of the shape of the parts or the nature of the material) the more the user must be able to adapt the robot to the environment.

Appendices

Appendix I
Matrix calculations in robotics: a summary

AI.1 Use of matrix calculations

1. In order to control a robot it is necessary to know the position (M) of the end effector in relation to the fixed base. M can be represented by its three components X_M, Y_M and Z_M in a fixed coordinate set linked to the base of the robot. The calculations involving matrices are used to deal with changes of coordinate sets implying movement, ie change in configuration as well as velocities, acceleration etc.

2. The robot has several limbs, each controlled by a motor. All these motors are activated simultaneously. Thus all the variables of articulation (θ_1, θ_2, θ_3 and θ_4) in Figure AI.1 must be considered together. The transformations performed on them can be expressed as *matrices*.

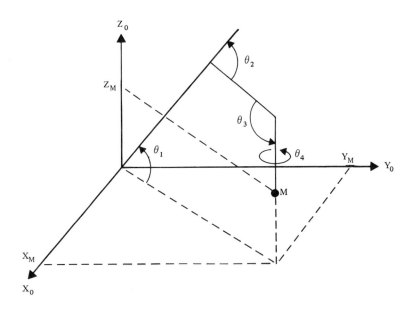

Figure AI.1. *Coordinates of M in the fixed set of coordinate axes $X_0 Y_0 Z_0$*

AI.2 Handling real term matrices: a summary

AI.2.1 DEFINITION

A matrix is expressed in a rectangular array of terms arranged in rows, which run horizontally, and columns, which run vertically.

Example:

$$[A] = \begin{pmatrix} a_1 & a_2 & a_3 & a_4 \\ b_1 & b_2 & b_3 & b_4 \\ c_1 & c_2 & c_3 & c_4 \end{pmatrix} \leftarrow \text{1st row} \qquad \text{(AI-1)}$$

$\uparrow$ 3rd column

AI.2.2 OPERATOR MATRIX

In this section, the matrix will be used as an operator (linear), ie it transforms starting values Y into target values X.

Example:

$$\begin{pmatrix} X_1 \\ X_2 \\ X_3 \end{pmatrix} = \begin{pmatrix} a_1 & a_2 & a_3 & a_4 \\ b_1 & b_2 & b_3 & b_4 \\ c_1 & c_2 & c_3 & c_4 \end{pmatrix} \begin{pmatrix} Y_1 \\ Y_2 \\ Y_3 \\ Y_4 \end{pmatrix} \qquad \text{(AI-2)}$$

$\uparrow$ target values constituting a vector of three components

$\uparrow$ starting values constituting a vector of four components

The operation performed by the matrix after expansion provides the following set of simultaneous equations:

$$X_1 = a_1 Y_1 + a_2 Y_2 + a_3 Y_3 + a_4 Y_4 \qquad \text{(AI-3)}$$

$$X_2 = b_1 Y_1 + b_2 Y_2 + b_3 Y_3 + b_4 Y_4 \qquad \text{(AI-4)}$$

$$X_3 = c_1 Y_1 + c_2 Y_2 + c_3 Y_3 + c_4 Y_4 \qquad \text{(AI-5)}$$

AI.2.3 MATRIX ADDITION $[A_1] + [A_2]$

Two matrices can be added term by term.

Example:

$$\begin{pmatrix} a_1 & a_2 \\ b_1 & b_2 \\ 0 & 1 \end{pmatrix} + \begin{pmatrix} c_1 & c_2 \\ d_1 & d_2 \\ e_1 & e_2 \end{pmatrix} = \begin{pmatrix} a_1 + c_1 & a_2 + c_2 \\ b_1 + d_1 & b_2 + d_2 \\ e_1 & e_2 + 1 \end{pmatrix} \qquad \text{(AI-6)}$$

AI.2.4 MATRIX MULTIPLICATION $[A_1 A_2] \neq [A_2 A_1]$

This is not commutative. The number of rows of one matrix must be equal to the number of columns of the other.

Example:

$$\begin{pmatrix} a_1 & a_2 & a_3 \\ b_1 & b_2 & b_3 \end{pmatrix} \begin{pmatrix} c_1 & c_2 \\ d_1 & d_2 \\ e_1 & e_2 \end{pmatrix} = \begin{pmatrix} a_1 c_1 + a_2 d_1 + a_3 e_1 & a_1 e_2 + a_2 d_2 + a_3 e_2 \\ b_1 c_1 + b_2 d_1 + b_3 e_1 & b_1 c_2 + b_2 d_2 + b_3 e_2 \end{pmatrix} \quad \text{(AI-7)}$$

AI.2.5 TRANSPOSED MATRIX $[A]^T$

The rows and columns are interchanged.

Example:

$$[A] = \begin{pmatrix} a_1 & a_2 & a_3 \\ b_1 & b_2 & b_3 \end{pmatrix} \quad \text{(AI-8)}$$

$$[A]^T = \begin{pmatrix} a_1 & b_1 \\ a_2 & b_2 \\ a_3 & b_3 \end{pmatrix} \quad \text{(AI-9)}$$

AI.2.6 MATRIX INVERSION $[A]^{-1}$

A. When the matrix is not square special techniques arising from the general theory of inverses must be applied.

B. When the matrix is square (ie the number of rows is equal to the number of columns):

(i) Its determinant is not zero — the standard method has three stages:

1. The matrix is *transposed*.
2. Each term of the transposed matrix is replaced by the corresponding *minor*.
3. Each final term is divided by the *determinant* of the initial matrix.

The following methods of calculation are used:
1. Determinant of $[A]$ equals Δ_A.

Example:

$$[A] = \begin{pmatrix} a_1 & a_2 & a_3 \\ b_1 & b_2 & b_3 \\ c_1 & c_2 & c_3 \end{pmatrix} \quad \Delta_A = \begin{vmatrix} a_1 & a_2 & a_3 \\ b_1 & b_2 & b_3 \\ c_1 & c_2 & c_3 \end{vmatrix} \quad \text{associated signs} \quad \begin{matrix} + & - & + \\ - & + & - \\ + & - & + \end{matrix} \quad \text{(AI-10)}$$

$$\Delta_A = a_1 \begin{vmatrix} b_2 & b_3 \\ c_2 & c_3 \end{vmatrix} - a_2 \begin{vmatrix} b_1 & b_3 \\ c_1 & c_3 \end{vmatrix} + a_3 \begin{vmatrix} b_1 & b_2 \\ c_1 & c_2 \end{vmatrix} \qquad (AI\text{-}11)$$

$$\Delta_A = a_1(b_2 c_3 - b_3 c_2) - a_2(b_1 c_3 - b_3 c_1) + a_3(b_1 c_2 - b_2 c_1) \qquad (AI\text{-}12)$$

The determinant can be found by developing along any row or column of A. Thus, the following can be written by developing from the second column of (A):

$$\Delta_A = -a_2 \begin{vmatrix} b_1 & b_3 \\ c_1 & c_3 \end{vmatrix} + b_2 \begin{vmatrix} a_1 & a_3 \\ c_1 & c_3 \end{vmatrix} - c_2 \begin{vmatrix} a_1 & a_3 \\ b_1 & b_3 \end{vmatrix} \qquad (AI\text{-}13)$$

2. The minor of a term equals the cofactor modified by the sign. In the preceding matrix:

$$\text{minor of } a_1 = \begin{vmatrix} b_2 & b_3 \\ c_2 & c_3 \end{vmatrix} = b_2 c_3 - b_3 c_2 \qquad (AI\text{-}14)$$

$$\text{minor of } b_3 = \begin{vmatrix} a_1 & a_2 \\ c_1 & c_2 \end{vmatrix} = -(a_1 c_2 - a_2 c_1) \qquad (AI\text{-}15)$$

etc.

3. Calculation of $[A]^{-1}$

Example:

$$[A] = \begin{pmatrix} a_1 & a_2 & a_3 \\ b_1 & b_2 & b_3 \\ c_1 & c_2 & c_3 \end{pmatrix} \qquad (AI\text{-}16)$$

$$[A]^T = \begin{pmatrix} a_1 & b_1 & c_1 \\ a_2 & b_2 & c_2 \\ a_3 & b_3 & c_3 \end{pmatrix} \qquad (AI\text{-}17)$$

$$[A]^T_{minor} = \begin{pmatrix} b_2 c_3 - b_3 c_2 & -(a_2 c_3 - a_3 c_2) & a_2 b_3 - a_3 b_2 \\ -(b_1 c_3 - b_3 c_1) & a_1 c_3 - a_3 c_1 & -(a_1 b_3 - a_3 b_1) \\ b_1 c_2 - b_2 c_1 & -(a_1 c_2 - a_2 c_1) & a_1 b_2 - a_2 b_1 \end{pmatrix}$$
$$(AI\text{-}18)$$

$$[A]^{-1} = \frac{[A]^T_{minor}}{\Delta_A} \qquad (AI\text{-}19)$$

Division by Δ_A means that every term of $[A]^T_{minor}$ is divided by this value.

(ii) The determinant of [A] is zero — the fact that the inverse of (A) cannot be calculated using the standard method — $\Delta_A = 0$ is important: the progression from starting values to resultant values can be performed in an infinite number of ways. The lack of supplementary information prevents the choice of one single method.

(iii) There are many techniques of matrix inversion. If a robot is to be controlled by digital computer, the standard method should be avoided, since it involves division by Δ_A. This is awkward, since computer precision is limited. Δ_A is rarely exactly equal to zero, and if this is the case the computer ceases to function. It will generally give an incorrect solution, which can lead to dangerous or unexpected robot behaviour.

AI.2.7 MULTIPLICATION OR DIVISION OF A MATRIX BY A CONSTANT ($\neq 0$)

Each term of the matrix is multiplied (divided) by the constant.

AI.2.8 UNIT MATRIX

An example of such a matrix would be:

$$(\mathbb{1}) = \begin{pmatrix} 1 & 0 & 0 \\ 0 & 1 & 0 \\ 0 & 0 & 1 \end{pmatrix} \qquad \text{(AI-20)}$$

AI.2.9 UNITARY MATRIX

This is a square matrix with real terms, which, when transposed, is equal to its inversion:

$$[A]^T = [A]^{-1} \qquad \text{(AI-21)}$$

This form of matrix will appear in all calculations of change of coordinate sets.

AI.2.10 DETERMINANT RANK OF A MATRIX

This is the size of the non-zero determinant of the largest minor it contains:

$$[A] = \begin{bmatrix} 1 & 0 & 0 \\ 2 & 1 & 1 \\ 3 & 2 & 2 \end{bmatrix} \quad \text{matrix of dimension 3 and rank 2} \qquad \text{(AI-22)}$$

Appendix II
Mathematial summary: transformation of coordinate sets

AII.1 Components of a vector in an orthogonal set of coordinates

An orthogonal set of coordinates is represented by three axes perpendicular to each other, each one with a unitary vector. These three vectors constitute a basic space:

$$\underline{e} = (e_x, e_y, e_z) \quad \text{(AII-1)}$$

where:
$$\underline{e_x} = (1, 0, 0) \quad \text{(AII-2)}$$
$$\underline{e_y} = (0, 1, 0) \quad \text{(AII-3)}$$
$$\underline{e_z} = (0, 0, 1) \quad \text{(AII-4)}$$

resolution axes: OX, OY, OZ. Note that:

1. A vector is expressed by its three components along OX, OY and OZ.
2. A component may also be considered as a vector, of which the projections on the two other axes are nil.

Example: e_z is the component of $\underline{e}$ on OZ; $\underline{e_z}$ is a unitary vector of component 0, 0, 1.

Consider the orthogonal set of coordinates (R) associated with origin e. Each point P in space can be defined with reference to the coordinate set (R) by the vector OP, which has three resolution axes OX, OY and OZ measured in units of e_x, e_y and e_z (see Figure AII.1).

$$\underline{OP}(R) = OX_P \underline{e_x} + OY_P \underline{e_y} + OZ_P \underline{e_z} \quad \text{(AII-5)}$$

or, expressed in its usual form:

$$P(R) = (X_P, Y_P, Z_P) \quad \text{(AII-6)}$$

Similarly, for another point Q:

$$Q(R) = (X_Q, Y_Q, Z_Q) \quad \text{(AII-7)}$$

and the vector $\underline{PQ} = \underline{V}_{PQ}$ has the following components in the coordinate set (R):

$$\underline{V}_{PQ}(R) = (X_Q - X_P, Y_Q - Y_P, Z_Q - Z_P) \quad \text{(AII-8)}$$

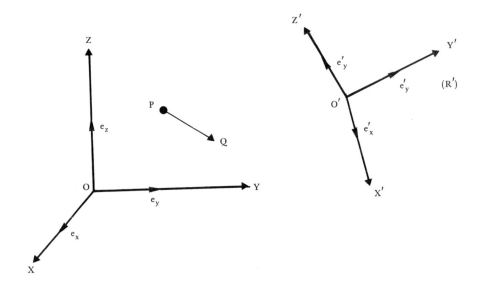

Figure AII.1.

AII.2 Transformation of coordinate set

It is now required to determine the components of V_{PQ} in a new coordinate set (R') with origin $\underline{e}' = (e'x, e'y, e'z)$.

Each component of $\underline{e}'$ can be written:

$$\underline{e}'_x(R) = \alpha_x \underline{e_x} + \beta_x \underline{e_y} + \gamma_x \underline{e_z} \qquad \text{(AII-9)}$$

$$\underline{e}'_y(R) = \alpha_y \underline{e_x} + \beta_y \underline{e_y} + \gamma_y \underline{e_z} \qquad \text{(AII-10)}$$

$$\underline{e}'_z(R) = \alpha_z \underline{e_x} + \beta_z \underline{e_y} + \gamma_z \underline{e_z} \qquad \text{(AII-11)}$$

α_x, β_x and γ_x are known as the cosine projectors of e'_x into (R). Taking into account the nature of the base vectors, the preceding system is identical to:

$$\underline{e}'(R) = \begin{pmatrix} \alpha_x & \beta_x & \gamma_x \\ \alpha_y & \beta_y & \gamma_y \\ \alpha_z & \beta_z & \gamma_z \end{pmatrix} \underline{e}(R) \qquad \text{(AII-12)}$$

but:

$$\underline{e}(R) \equiv \underline{e}'(R') = (1, 1, 1) \qquad \text{(AII-13)}$$

Thus:

$$\underline{e}'(R) = \begin{pmatrix} \alpha_x & \beta_x & \gamma_x \\ \alpha_y & \beta_y & \gamma_y \\ \alpha_z & \beta_z & \gamma_z \end{pmatrix} \underline{e}'(R') \qquad \text{(AII-14)}$$

- $\underline{e}'(R)$: basis of (R') expressed in (R)
- matrix: coordinate transformation matrix $M_{R'}^{R}$
- $\underline{e}'(R')$: basis of (R') expressed in (R')

The basis is a vector like any other. Thus:

$$\underline{V}(R) = M_{R'}^{R} \, \underline{V}(R') \qquad \text{(AII-15)}$$

- $\underline{V}(R)$: vector expressed in (R)
- $M_{R'}^{R}$: coordinate transformation matrix
- $\underline{V}(R')$: vector expressed in (R')

And for the vector $\underline{V}_{PQ}$:

$$\underline{V}_{PQ}(R) = M_{R'}^{R} \, \underline{V}_{PQ}(R') \qquad \text{(AII-16)}$$

or

$$\underline{V}_{PQ}(R') = [M_{R'}^{R}]^{-1} \, \underline{V}_{PQ}(R) \qquad \text{(AII-17)}$$

Thus, the matrix of transformation of the coordinate set is obtained from the cosine projectors of the new origin relative to the old origin.

AII.3 Specific examples useful for modelling and control of robots

AII.3.1 (R') IS OBTAINED BY ROTATION THROUGH ANGLE θ OF COORDINATE SET (R) ABOUT THE Z AXIS

$$\underline{e}'_x(R) = \underline{e}_x \cos\theta + \underline{e}_y \sin\theta \qquad \text{(AII-18)}$$

$$\underline{e}'_y(R) = \underline{e}_x(-\sin\theta) + \underline{e}_y \cos\theta \qquad \text{(AII-19)}$$

$$\underline{e}'_z(R) = \underline{e}_z \qquad \text{(AII-20)}$$

and:

$$M_{R'}^{R}(Z) = \begin{pmatrix} \cos\theta & \sin\theta & 0 \\ -\sin\theta & \cos\theta & 0 \\ 0 & 0 & 1 \end{pmatrix} \qquad \text{(AII-21)}$$

Practical note: To find $M_{R'}^{R}$, the projections of the unitary vectors of (R'), which is mobile, on the axes of (R) which remain fixed, must be considered.

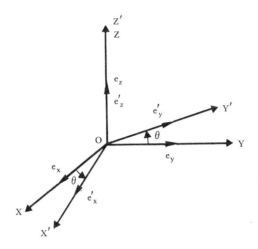

Figure AII.2. *Rotation about OZ*

AII.3.2 (R′) IS OBTAINED BY THE ROTATION OF THE SET OF COORDINATE AXES (R) THROUGH AN ANGLE OF θ ABOUT Y

Using the same method as before:

$$M_{R'}^{R}(X) = \begin{pmatrix} \cos\theta & 0 & -\sin\theta \\ 0 & 1 & 0 \\ \sin\theta & 0 & \cos\theta \end{pmatrix} \quad \text{(AII-22)}$$

AII.3.3 (R′) IS OBTAINED BY THE ROTATION OF THE SET OF COORDINATE AXES (R) THROUGH AN ANGLE θ ABOUT X

$$M_{R'}^{R}(Y) = \begin{pmatrix} 1 & 0 & 0 \\ 0 & \cos\theta & \sin\theta \\ 0 & -\sin\theta & \cos\theta \end{pmatrix} \quad \text{(AII-23)}$$

AII.3.4 (R′) IS OBTAINED BY ANY TRANSLATION OF (R)

The $\underline{e}'_x$, $\underline{e}'_y$ and $\underline{e}'_z$ remain parallel to $\underline{e}_x$, $\underline{e}_y$ and $\underline{e}_z$ (it could also be said that $\overline{\theta = 0}$):

$$M_{R'}^{R}(T) = \begin{pmatrix} 1 & 0 & 0 \\ 0 & 1 & 0 \\ 0 & 0 & 1 \end{pmatrix} = (\mathbb{1}) \quad \text{(AII-24)}$$

$$\underline{V}(R) = \underline{V}(R') \quad \text{(AII-25)}$$

AII.4 Inverse transformation

The equation:
$$V(R) = M_{R'}^{R} V(R') \tag{AII-26}$$
is inverted thus:
$$\underline{V}(R') = [M_{R'}^{R}]^{-1} \underline{V}(R) \tag{AII-27}$$

The matrix $M_{R'}^{R}$, which is also called a *matrix of coordinate transformation*, is unitary. The result is, as already seen:
$$[M_{R}^{R'}(Z)] = [M_{R'}^{R}]^{T}$$
which is written thus: $[M_{R}^{R'}]$.

Example:
$$[M_{R}^{R'}(Z)] = [M_{R'}^{R}(Z)]^{T} = \begin{pmatrix} \cos\theta & -\sin\theta & 0 \\ \sin\theta & \cos\theta & 0 \\ 0 & 0 & 1 \end{pmatrix} \tag{AII-28}$$

So generally:
$$\underline{V}(R') = M_{R}^{R'} \underline{V}(R) = [M_{R'}^{R}]^{T} \underline{V}(R) \tag{AII-29}$$

Appendix III

Summary of the principles of hydraulic flow

AIII.1 Definitions and equations

The general equation of flow depends on the principle of conservation of mass in any part of the hydraulic system in which the pressure can be considered uniform at a given time. Expressed in general terms, the variation in mass M of the fluid contained between two sections is equal to the difference between the incoming and outgoing flows D_e and D_s expressed as a mass.

During a time interval T, a mass M_e enters and a mass M_s leaves a section of the flow circuit (see Figure AIII.1).

Figure AIII.1.

The incoming flow expressed as a mass is:

$$D_e = M_e/\Delta T \tag{AIII-1}$$

The outgoing flow is:

$$D_s = M_s/\Delta T \tag{AIII-2}$$

The variation of the continuous mass in the section of circuit during the time interval ΔT is thus:

$$\frac{\Delta M}{\Delta T} = D_e - D_s \tag{AIII-3}$$

or, in the limit:

$$\frac{dM}{dt} = D_e - D_s \tag{AIII-4}$$

Equation (AIII-4) can be expressed as a volume instead of as a mass,

where: $M = \rho V$ \hfill (AIII-5)

and: $D = \rho Q$ \hfill (AIII-6)

where ρ is the specific gravity of the fluid, V is the volume of the fluid element and Q the volumetric flow. Equation (AIII-4) becomes:

$$\frac{dV}{dt} + \frac{V}{\rho}\frac{d\rho}{dt} = Q_e - Q_s \qquad \text{(AIII-7)}$$

This is the basic formula describing hydraulic flow. Consider the first two terms in equation (AIII-7).

1. dV/dt represents flow caused by volumetric change in the part of the flow system under consideration. This variation may have one of several possible causes:
    — It can be deliberately caused by moving a piston-like mobile component. Its mode of calculation will depend on its cause.
    — It can be caused by involuntary elastic deformation in the system. In most cases this variation in volume is proportional to the variation in pressure. Contrary to what might be expected, this deformation may have significant effects, especially in the flexible connecting tubes, which should be avoided if high-performance standards are required.
2. The second term $(V/\rho)(d\rho/dt)$ represents a volume flow due to the compressibility of the fluid used. Fluids are generally compressible to such a small degree that ρ can be considered to be constant in the range of pressures likely to be encountered. The rate of variation, however, is often considerable, so that the term $(V/\rho)(d\rho/dt)$ is not always negligible.

In practice, for a known range of temperature and pressure, a coefficient of compressibility B is defined:

$$\frac{\Delta\rho}{\rho} = \frac{\Delta P}{B} \qquad \text{(AIII-8)}$$

where P is the pressure, so that the compressibility term takes the form: $(V/B)(dP/dt)$.

Appendix IV

Direct current motors

AIV.1 Working principles

All servocontrol electric motors are based on the existence of the Laplace force. A conducting element $d\ell$, through which a current flows, is placed in a magnetic field B, and is subject to a force:

$$\underline{dF} = I\underline{d\ell}\ \underline{B} \qquad (AIV-1)$$

Continuous current motors are characterized by the fact that the field direction is fixed. This field can be formed in two ways:

— by a permanent magnet;
— by an electromagnetic circuit.

Suppose that the conductor in formed by a single coil as shown in Figure AIV.1. If the coil is supplied with a current, flowing in the direction shown, parts AB and CD are subjected to two equal and opposite Laplace forces, which constitute torque. The coil then rotates about its axis.

In practice, the robot has many windings. To improve the perform-

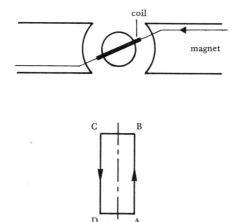

Figure AIV.1. *Principle of the direct current (d.c.) motor*

ance of these motors, inertia must be reduced to a minimum. Robots with motors, up to a power of several kilowatts, are formed of several layers of strip conductors, superimposed and insulated by sheets of epoxy glass. This constitutes a coil wound in series. In servocontrol systems, these motors are used in an arrangement of 'separate excitation', that is, the electromagnetic inductor and the armature are supplied separately. Two methods of control are possible:

— induction control;
— armature control.

AIV.2 Motor with induction control

Figure AIV.2 shows the diagram usually used to represent this kind of drive. Under this type of control, the inductor is supplied with a variable

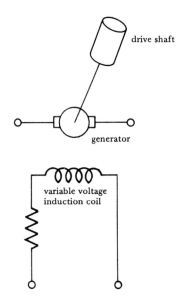

Figure AIV.2. *Motor with induction control*

voltage, which forms the motor control, whereas the armature is supplied with a constant voltage. Under normal conditions the characteristics of torque versus speed of rotation can be observed. These are shown in Figure AIV.3, as a collection of straight lines.

The deviation from linearity which occurs at high speed is a result of non-linear physical phenomena. Nonetheless, the effective torque is not sensitive to this change over a large range of speeds in normal use.

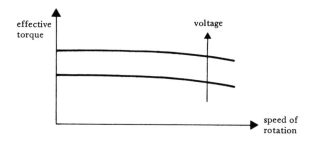

Figure AIV.3. *Torque and speed of rotation characteristics for the induction control motor*

AIV.3 Motor with armature control

This type of control is usually represented as shown in Figure AIV.4. In this case the inductor is supplied at constant current and thus produces a constant flux like a permanent magnet. The armature is set at a variable voltage, which forms the control variable. The shapes of the

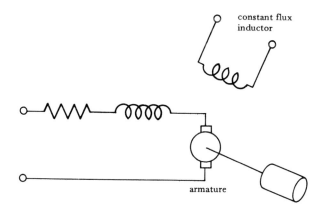

Figure AIV.4. *Motor with armature control*

characteristic curves described in use are different to those described by induction controlled motors, as seen in Figure AIV.5.

This type of control is used widely. The constant flux is produced by permanent magnets of cast alloys, magnetized during assembly, which are stable over a temperature range.

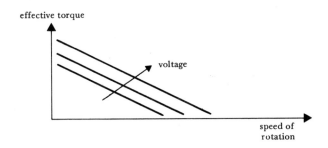

Figure AIV.5. *Torque and speed of rotation characteristics for the armature control direct current motor*

Appendix V
The dynamic model

AV.1 Problems associated with dynamic control

The problems that arise in the process of model formulation are many and varied, so only a limited number of parameters can be corrected at once. It is assumed that the robot is very rigid (no elasticity), and the articulations are perfect (no friction), with simple geometry (no movement relative to the rotational centres). Equations of movement must be established in relation to the forces present. What are these forces?

1. gravity;
2. forces that arise from the motors and apply to the articulations;
3. forces that arise from the movement of masses:
   α: inertial forces (proportional to acceleration);
   β: centrifugal forces (proportional to velocity squared);
   γ: forces arising from the coupling between segments, known as the Coriolis forces (proportional to the products of the velocities of various articulations).

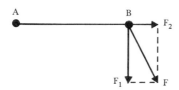

Figure AV.1. *Torque on A: $F_1 \times AB$*

When studying rotational articulations (see Figure AV.1) it is important to know what happens at A. At A there is a torque which is the product of force F (projection of F on a normal to AB) and the length AB. In order to assess the effect of forces on a robot, methods known as *formalisms* must be applied. The most commonly used is the *Lagrange method*. The method will not be explained here, but certain elements of the result should nonetheless be carefully studied. The model obtained can be written in matrix form:

$$[A(\underline{\Theta})](\underline{\ddot{\Theta}}) + [B(\underline{\Theta})]\underline{\dot{\Theta}}^2 + [C(\underline{\Theta})]\underline{\dot{\Theta}\dot{\Theta}} = Q(\underline{\Theta}) + \underline{\Gamma} \quad \text{(AV-1)}$$
$$\quad N\times N \qquad\qquad N\times N \qquad\quad N\times C_N^2$$

This is the condensed form of a system of differential equations of the second order, which are coupled and non-linear.

N is the number of a robot's articulations:

$$C_N^2 = \frac{N!}{2!N-2!} = \frac{N(N-1)}{2} \quad \text{(AV-2)}$$

$\underline{\ddot{\Theta}}$ is the vector of articulation acceleration:

$$\underline{\ddot{\Theta}} = (\ddot{\theta}_1, \ddot{\theta}_2, \ldots \ddot{\theta}_N)^T \quad (\ddot{\theta} = d^2\theta/dt^2) \quad \text{(AV-3)}$$

$\underline{\dot{\Theta}}^2$ is the vector of the square of articulation velocity:

$$\underline{\dot{\Theta}}^2 = (\dot{\theta}_1^2, \dot{\theta}_2^2 \ldots \dot{\theta}_N^2)^T \quad (\dot{\theta} = d\theta/dt) \quad \text{(AV-4)}$$

$\underline{\dot{\Theta}\dot{\Theta}}$ is the vector of the product of the articulation velocities:

$$\underline{\dot{\Theta}\dot{\Theta}} = (\dot{\theta}_1\dot{\theta}_2, \dot{\theta}_1\dot{\theta}_3, \ldots \dot{\theta}_1\dot{\theta}_N, \dot{\theta}_2\dot{\theta}_3, \ldots \dot{\theta}_2\dot{\theta}_N,$$
$$\ldots \dot{\theta}_{N-1}\dot{\theta}_N)^T \quad \text{(AV-5)}$$

$[A(\underline{\Theta})]$ $[B(\underline{\Theta})]$ $[C(\underline{\Theta})]$ are matrices in which each term depends on the values of the articulations. These terms are called the *dynamic coefficients* of the robot, and are highly complex. For a robot with six DOF, for example, the first term of the matrix $[A(\underline{\Theta})]$, which will be called α_{11}, is written on about twenty lines (in this example there are 162, but all different, and not all so complicated).

$[A(\underline{\Theta})]\underline{\ddot{\Theta}}$ represents the inertial forces or torques acting on the robot.
$[B(\underline{\Theta})]\underline{\dot{\Theta}}^2$ represents the centrifugal forces or torques.
$[C(\underline{\Theta})]\underline{\dot{\Theta}\dot{\Theta}}$ represents the Coriolis forces.
$Q(\underline{\Theta})$ represents the torque because of gravity.
$\underline{\Gamma}$ represents the external torques, which in this example are reduced to the torque arising from the motors and applied to the articulations. It should be noted that:

1. Although very few parameters are taken into consideration in comparison with reality, the dynamic model is extremely complicated and non-linear, and this will affect control.
2. The velocity and acceleration of the robot can be seen, which is a great advantage.

AV.2 Dynamic control

This is carried out by inverting the model in equation (AV-1), which means that if $\underline{X}(t)$, ie the position and orientation of the end effector in coordinate set (R_0) for example, is known, and it is possible to move from $X(t)$ to $\theta(t)$ (resolvable robot, with kinematic model), it is necessary to find what $\Gamma(t)$ should be applied to the articulations to make the robot perform $\theta(t)$ and thus $\underline{X}(t)$. This is a complex problem because:

1. the equations are non-linear, and non-linear control is difficult;
2. the equations are very complicated, and thus the calculations are long. Computers are not fast enough to perform the calculations in real time.

Even using solutions to this problem which have been developed in advanced research does not allow rapid enough control. Moreover this model takes into account only a few parameters. If the effects of friction are considered, for example, the calculation becomes enormously complex.

AV.3 Effects of gravitational force

If the forces listed in the left side of the equation (AV-1) are linked to the movement (velocity and acceleration) the result does not equal the elements on the right side of the equation. Thus, when the robot is at rest in a given configuration, equation (AV-1) can be expressed as:

$$\underline{Q}(\Theta) + \underline{\Gamma} = \underline{0} \qquad \text{(AV-6)}$$

which is the same as:

$$\begin{aligned} Q_1(\theta_1, \theta_2 \ldots \theta_N) + \Gamma_1 &= 0 \\ Q_2(\theta_1, \theta_2 \ldots \theta_N) + \Gamma_2 &= 0 \\ &\vdots \\ Q_N(\theta_1, \theta_2 \ldots \theta_N) + \Gamma_N &= 0 \end{aligned} \qquad \text{(AV-7)}$$

$Q_1, Q_2, \ldots Q_N$ are the gravitational torques acting on articulations $1, 2, \ldots N$, and $\Gamma_1, \Gamma_2, \ldots \Gamma_N$ are the torques arising from the motors and acting upon articulations $1, 2, \ldots N$. In practice, a new problem arises with these parameters, no matter which method of control is used. To counteract the gravitational pull, the motors have to work to overcome torques $\Gamma_1, \Gamma_2, \ldots \Gamma_N$. There are two possibilities:

1. The robot is irreversible (with ratchet transmission, for example). This means that if the motors stop, the mechanical transmission prevents the articulations from slipping back.
2. The robot is reversible (if pressure is applied to the end effector, it affects the motors). In this example the pressure applied to the

end effector can be deduced by measuring the drive current. However if the motors are switched off, the force of gravity will cause the articulations to drop. To compensate for this, and to maintain the advantages of reversibility, most robots are equipped with mechanical counterbalances. These consist of weights which move with the robot segments, and allow the robot to maintain its configuration when the motors are switched off. (See, for example, Figure AV.2.)

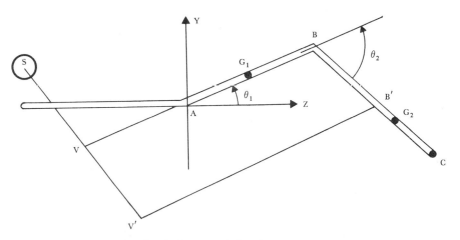

Figure AV.2. *Mechanical counterbalancing using the parallelogram*

As an example, consider a simplified robot OABC, which only moves in the plane YAZ. AY represents the vertical. G_1 and G_2 are the mass centres of AB and BC, of mass M_1 and M_2. The parallelogram $VV'BB'$ is constructed such that VA, VV' and VB' are of negligible mass, and the mass of S is M. The ordinates (heights) of the mass centres are:

$$YS = VS \cdot S(1+2) - AV \cdot S1 \qquad \text{(AV-8)}$$

$$YG_2 = 1/2 AB \cdot S1 \qquad \text{(AV-9)}$$

$$YG_3 = AB \cdot S1 - 1/2 BC \cdot S(1+2) \qquad \text{(AV-10)}$$

The torques caused by gravity in A and B can be found using the principle of potential work (see Figure AV.3). According to this principle, the work of torque Q exerted on O by a small rotation $\Delta\alpha$ is equal to the work force F in the vertical movement it causes:

$$Q \cdot \Delta\alpha = F \cdot \Delta y \qquad \text{(AV-11)}$$

from whence:

$$Q = F \frac{\Delta y}{\Delta \alpha} \cong F \frac{\partial y}{\partial \alpha} \qquad \text{(AV-12)}$$

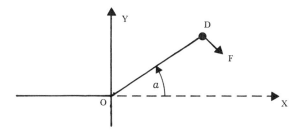

Figure AV.3. *Principle of potential work: the torque is created by force F*

This principle can be applied to the example:

$$Q_A = g\left(M\frac{\partial Y_S}{\partial \theta_1} + M_1 \frac{\partial Y_{G_1}}{\partial \theta_1} + M_2 \frac{\partial Y_{G_2}}{\partial \theta_1}\right) \quad \text{(AV-13)}$$

$$Q_B = g\left(M\frac{\partial Y_S}{\partial \theta_2} + M_1 \frac{\partial Y_{G_1}}{\partial \theta_2} + M_2 \frac{\partial Y_{G_2}}{\partial \theta_2}\right) \quad \text{(AV-14)}$$

g is the acceleration due to gravity. The torques Q_A and Q_B can be cancelled out if:

$$M \cdot VS = M_2 \cdot BC/2 \quad \text{(AV-15)}$$

$$M \cdot AV = M_1 \cdot AB/2 + M_2 \cdot AB \quad \text{(AV-16)}$$

VS and AV can be calculated from equations (AV-15) and (AV-16) if M is imposed. Whatever the values of θ_1 and θ_2 it will still be balanced.

In reversible robots the use of mechanical counterbalancing (of the parallelogram type) should be introduced to avoid problems when the motors stop. If g = 0 (space robots) this counterbalancing is not necessary.

Bibliography

Historical
Cohen, J. *Human Robots and Myth in Science* Allen and Unwin, London, 1966.
Geduld, H.; Gottesman, R. *Robots, Robots, Robots* New York Graphic Society, New York, 1978.

Introduction
Engelberger, J.F. *Robotics in Practice* Kogan Page, London, 1982.
Industrial Robots Volume 1, Society of Manufacturing Engineers, 1981.
Industrial Robots Volume 2, Society of Manufacturing Engineers, 1981.
Proceedings of the Royal Society of Arts (in preparation).

Detailed
Aleksander, I. *Artificial Vision for Robots* Kogan Page, London, 1983.
Coiffet, P. *Modelling and Control* Robot Technology, Volume 1, Kogan Page, London, 1983.
Coiffet, P. *Interaction with the Environment* Robot Technology, Volume 2, Kogan Page, London, 1983.
Lhote, F.; Kauffmann, J.; Andre, P.; Taillard, J. *Robot Components* Robot Technology, Volume 4, Kogan Page, London (in press).
Liegeois, A. *Robot Performance Evaluation and Computer-aided Design* Robot Technology, Volume 7, Kogan Page, London (in preparation).
Parent M. *Robotic Languages and Progamming Methods* Robot Technology, Volume 5, Kogan Page, London (in preparation).
Prajoux, R.; Farreny, H.; Ghallab, M. *Decision Autonomy and Artificial Intelligence* Robot Technology, Volume 6, Kogan Page, London (in preparation).
Vertut, J.; Coiffet, P. *Teleoperations* Robot Technology, Volume 3, Kogan Page, London (in preparation).

Advanced research robots
Nitzan, D. *Flexible Automation Programs at SRI, Proceedings of the 1979 Joint Automatic Control Conference* IEEE, New York, 1979.

Developments in computing
Large, P. *The Micro Revolution* Fontana, London, 1980.

Miscellaneous
Annan, D. *Robot, the Mechanical Monster* Bounty Books, New York, 1976.

Index

absolute encoder, 113
  disk, 114
acceleration sensors, 119
ACMA robot, 152, 154, 158
  P7 robot, 158
active training, 138
actuators, 18, 83-105
  electrical, 83, 91
  hydraulic, 83
  pneumatic, 83-4
actuator servocontrol, 67-82
AFRI classification, 22
aids for the handicapped, 14
AKR robot, 158
AMS, functional representation of, 25-7
analog control servo-system, 80
analog-digital converter (ADC), 73, 130
Andromat robot, 153
angular movement sensors, 109
arm, robot, 26
  structures in use, 29
articulated mechanical chain
  associated coordinated sets, 44
  representation of, 43
  simplified representation, 44
articulated mechanical system (AMS), 18
artificial intelligence (AI)
  control of robot, 39
  languages, 145
artificial skin, 123
artificial vision, 121
ASEA IRb-6 robot, 158-9
attainable volume, 148
auto-adaptability to environment, 18
automation, development of, 11
autonomous robot, 13
AUTOPASS language, 145
AW7 (General Electric) robot, 155

BASIC language, 132
bell motor, 92-3
binary code, encoder, 114
Black locus, 78
Bond locus, 78
BUILD language, 145
buses, 133

central microprocessing unit control, 129
central processing unit, 133
charge coupled device (CCD), 127
  cameras, 128
closed-loop mode, actuator servocontrol, 69
compensation, 71-2, 79
compensation network, 100, 101
  proportional and derivative, 101
  proportional and integral, 102
  tachometric, 103
  three term, 102
complementary cube test, 152
computer-aided design (CAD), 12
computer-aided manufacture (CAM), 12
computer control of robot, 129-35
computer, in robot, 18
computer, role of, 20
control, direct, 68
  geometrical or positional, 51-6
  kinematic, 57
  languages, 145
  open-loop mode, 68
control of robot
  artificial intelligence, 39
  control mode, 39
  servo-system control, 40
conventional representation of robot, 45
converter
  microprocessor compatible, 130
  synchro/resolver, 130
  voltage-frequency, 130-1
coordinate sets, 173
  inverse transformation, 177
  transformation of, 174

coordinate task set, 43
counterbalance, 168

d.c. motors, 91
degrees of freedom (DOF), 26
   and mobility, 35
   rigid object, 33
   robot, 34
   rotating drill, 34
   rotational, 26, 33
   specific, 35
   sphere, 34
   translational, 26, 33
degrees of mobility, 35-6
depalletization, 153-4
determinant, 169
differential amplifier, use in servocontrol, 68
differential transformer, matrix sensor, 123
digital-analog converter (DAC), 73
digital control, servo-systems, 80
direct current (d.c.) motors, 91-5, 181-4
   armature control, 183-4
   induction control, 182
   working principle, 181-2
disk motor, 92, 94
double gripper end effector, 155
dynamic control of robot, 180
   problems, 185
   use of models, 65
dynamic precision, 149
dynamic random access memory (RAM), 134

economic performance of robot, 150
electrically powered actuators, 83-4
electro-hydraulic systems, 90-1
EMILY language, 145
encoders
   absolute, 113
   incremental, 111
   optical, 111
end effector, 26
   possible orientation of, 148
   structure of, 29-31
environment of robot, 19-20
ergonomics, 141
exploration, use of robots in, 13
extensiometric gauges, 111
   measurement, principles of, 115
external sensors, 107, 121-8

feedback, in servocontrol, 69

feeding and unloading machine tools, 153
fixed coordinate set, 43
form recognition, 123
formalisms, 185
FORTRAN language, 132
four-track spool distributor, 86

geometrical control, 51-6
geometrical model of robot, 43-51
   calculations, 45-51
graphical representation of robot, 27-9
gravitational forces, effects of, 187
gripper, 159, 160-1

high-level languages, 145
high-temperature handling, 153
human control of robot, 150
hydraulic actuator, 83
hydraulic flow, definitions and equations, 179-80
hydraulic servocontrol system, 89

important configurations, 140
incremental encoder, 111
   circuit of, 113
   disk, 112
individual analog sensors, 122
industrial robot, 21
infra-red sensors, 126
input-output circuits, 133
instrumentated platform, 125
internal sensing, 18
   sensors, 107-20
isolated binary contact sensors, 122

JIRA classification, 22
joystick control, 141

kinematic control, 64-5
kinematic model, 57-8

Lagrange method, 185
LAMA language, 145
language, assembly, 132
   control, 145
   evolved or high-level, 132
   high-level, 145
Laplace force, 181
Linc-man robot, 155
linear piston device, 85-7
   differential, 87
   double-action, 87
   single-action, 86

LM language, 145
load transportation, 153

matrices, 167, 171
    addition, 168
    definition, 168
    determinant rank, 171
    inversion, 169
    matrix calculation, 167
    multiplication, 169
    transposition, 169
    unitary, 171
matrix sensors, 123
mechanical articulation, representation of, 28
mechanical counterbalancing, 188
memory, 133-4
    RAM, 134
    ROM, 134
microcomputer, choice of, 134-5
    microprocessor compatible converter, 130
    program, 131-4
    structure of, 133
Modan PPH 50X robot, 163
modelling and control, transformation of coordinates, 175-7
motors
    armature controlled, 99-100
    bell armature, 92-3
    direct current (d.c.), 91
    disk, 92, 94
    induction controlled, 97
    servocontrolled, 96
    standard, 92
    stepping, 95
movement sensors, 107

Nyquist locus, 78

optical encoder, 111
ortheses, 14

paint spraygun tool, 163
painting robots, 158
palletization, 153
PASCAL language, 132
passive training, 137
pay load, 148
performance standard, testing of, 151
permanent continuous recording, 138
permanent discrete recording, 138
piezoelectric sensors, 119

piston actuators, 87
    differential, 87
    double-action, 87
    flapper, 88
    linear, 85-6
    rotary, 85, 87
    single-action, 86
plan generators, 145
pneumatic actuators, 83
    basic principles, 84
    systems, 85
pneumatic valve distributor device, 85
positional control, 51-6
    advantages and disadvantages, 54-6
    algorithm, 51
positional precision, 148
positional sensor, 107
positional servocontrol, 100
positional servo-systems, 73
    compensated, 74
potential work, 189
potentiometer
    coiled track, 107
    hybrid, 108
    plastic track, 108
    rectilinear, 107
potentiometer sensor, 109
power-to-weight ratio, 84
printed circuit assembly board, 156
processor, 37
program, microprocessor, 131-3
prostheses, 14
proximity sensor, 125-6
Puma robot, 156

rack and pinion system, 85
recording
    permanent continuous, 138
    permanent discrete, 138
    specific configuration, 139
    with passive robot, 139
redundant equations, 50-1
reliability of robot, 150
Remote Centre Compliance (RCC), 159, 162
resolvability, 51-2
response time of motor, 68
RIA classification, 22
robot
    arm, 26
    attainable volume, 148
    autonomous, 13
    classification systems, 22

robot (*continued*)
  computer control of, 18, 129
  control, levels of, 39-40
    variables to be handled, 37-8
  definition, 17
  degrees of freedom, 34
  dynamic control, 65, 180, 185
  economic performance, 150
  end effector, 26, 148, 155
  environment, 19-20
  function, representation of, 19, 25-7
  general structure, 18
  generations, 22
  geometrical model, 43-51
  graphical representation, 27-9
  gripper, 159, 160-1
  human control of, 150
  industrial, 21
  interactive, 12
  introduction of, 11
  origin of term, 17
  performance, 147
  processor, 37
  properties, 17 *et seq*
  reliability, 150
  representation, 25 *et seq*
  servocontrol, 40
  singularity, 61
  training, 137
    active, 138
    passive, 137
  trajectory, 137, 142
  uses and markets, 23-4
  vehicle, 26
  velocity, 149
  wrist, 124-5
robotics, applications of, 12-13
  development of, 11
rotating piston device, 85
rotational degrees of freedom, 26, 33

S6 (General Electric) robot, 158
scene analysis, 21
sensors, 18
  acceleration, 18
  angular movement, 109
  external, 107, 121 *et seq*
  individual analog, 122
  infra-red, 126
  internal, 107-20
  isolated binary control, 122
  matrix, 123
  movement, 107
  piezoelectric, 119
  positional, 107
  potentiometer, 107, 109
  proprioceptive, 107, 109
  proximity, 125 *et seq*
  servo-return, 119
  speed, 115
  stress, 116 *et seq*
  tactile, 122 *et seq*
  ultrasonic, 125
  variable transformer, 108
  visual, 126 *et seq*
sensor control of motor, 68
sensory feedback system, 14
servocontrol
  accuracy, 70
  actuator, 67 *et seq*
  closed-loop mode, 69
  compensation, 71-2
  hydraulic, 89-90
  servo-return, 119
  stability, 71
servocontrolled hydraulic system, 89-90
servocontrolled motor, 96 *et seq*
servo-return sensor, 119
servo-system, control of motor, 40
  digital control algorithm, 82
  frequency response, 75
  mathematical behaviour, 75
  open-loop, 77
  qualitative analysis of, 77
  stability, 77
  types of, 73
  use in robotics, 79-82
servovalve, hydraulic, 88-9
shoe carding, 158
SIGLA language, 145
singular configuration, 61
  examples, 62
singularity, 61-4
  physical meaning of, 61
small unit assembly work, 155
speed control model, 57
speed sensors, 115
  tachometric generators, 116
spool distributor, 85, 88
standard motors, 92
static read only memory (ROM), 134
steering mechanism (SM), 140
step-down gears, 97, 105
stepping motors, 95
strain gauge, 116-17, 124
stress sensor, 116 *et seq*, 124
STRIPS language, 145

structures, representation of (P, R convention), 29
synchronized capability of robot, 151
synchro/resolver converter, 130

tachometric generator, as speed sensor, 116
task axes, 37
task performance, 147
teleoperation, 13-14
teletheses, 14
TRAFLA robot, 158
training, robot, 137
   active, 138
   manual control in, 140
   passive, 137
trajectory calculation, 142-3
   in task space, 144
trajectory generation, 137, 142-4
   recording, 137
transfer function, 96
translational degrees of freedom, 26, 33
transmission system, 18, 103, 104
   choice of, 104-5

ultrasonic sensors, 125

VAL language, 145
valve distributor, 85
variable transformer sensors, 108, 110
variational control, 60
variational model, 57, 60
vector components, 173-5
vehicle, 26
velocity control, 142
velocity of robot, 149
Vidicon tube, 128
visual sensors, 126 *et seq*
voltage frequency converter, 130

Walkman assembly, 156
WAVE language, 145
Welding, 155-7
Westinghouse W7000 robot, 155